T0195300

Beyond Sputnik and the Space Race

Beyond Sputnik
and the Space Race

The Origins of Global Satellite Communications

HUGH R. SLOTTEN

Johns Hopkins University Press

Baltimore

© 2022 Johns Hopkins University Press
All rights reserved. Published 2022
Printed in the United States of America on acid-free paper

2 4 6 8 9 7 5 3 1

Johns Hopkins University Press
2715 North Charles Street
Baltimore, Maryland 21218-4363
www.press.jhu.edu

Library of Congress Cataloging-in-Publication Data
Names: Slotten, Hugh Richard, author.
Title: Beyond Sputnik and the space race : the origins of global satellite
communications / Hugh R. Slotten.
Description: Baltimore : Johns Hopkins University Press, 2022. |
Includes bibliographical references and index.
Identifiers: LCCN 2021011718 | ISBN 9781421441221 (hardcover) |
ISBN 9781421441238 (ebook)
Subjects: LCSH: Artificial satellites in telecommunication—United States—History. |
Astronautics—United States—History. | Space race—United States—History.
Classification: LCC TK5104 .S56 2022 | DDC 621.382/50973—dc23
LC record available at https://lccn.loc.gov/2021011718

A catalog record for this book is available from the British Library.

*Special discounts are available for bulk purchases of this book. For more information,
please contact Special Sales at specialsales@jh.edu.*

CONTENTS

Many people helped with this book; unfortunately, it isn't possible to thank everyone by name. But in addition to the staff at the Johns Hopkins University Press, I especially would like to acknowledge the extremely valuable assistance provided by staff at the Smithsonian Institution Air and Space Museum and the NASA History Office, as well as James Schwoch at Northwestern University, Richard R. John at Columbia University, and the anonymous referees who read the manuscript.

I am deeply grateful for fellowships and grants I received to support this project from the Smithsonian Institution Air and Space Museum (Charles A. Lindbergh Chair in Aerospace History and the Aviation Space Writers Foundation Award); the NASA History Office (AHA/NASA Aerospace History Fellowship); the University of Otago (Otago Research Grant and a Research and Study Leave Grant); and the National Science Foundation (NSF Research Grant). I also would like to thank the Department of the History of Science and Technology at Johns Hopkins University, where I spent a very productive year working on this project as a visiting scholar.

Beyond Sputnik and the Space Race

Introduction

On July 20, 1969, a remarkable percentage of the Earth's population watched a live television broadcast showing NASA astronaut Neil Armstrong step off the Apollo 11 lunar module to walk on the moon and speak the famous words, "That's one small step for man, one giant leap for mankind." NASA estimated that 20 percent of the planet's population (nearly 650 million people) viewed the moon landing live.[1] Underscoring the crucial role of communications satellites in this broadcast, Lisa Parks has called this and similar global media events "satellite spectaculars."[2] Live television from the moon was made possible by the first global satellite communications system, the International Telecommunications Satellite Organization (Intelsat). Receiving stations in Australia picked up the television signal from the moon and relayed it to NASA's Manned Spaceflight Center in Houston using the *Intelsat III* satellite over the Pacific Ocean. The first Intelsat satellite, *Intelsat I* (also known as *Early Bird*), which had been placed on a reserve status in 1968, was reactivated to assist with the necessary global relay links.[3]

A satellite communications distribution system functions by sending television video—as well as telephone calls, telegraph messages, and electronic data of all kinds—from a radio transmitter on a ground station to a radio receiver on a satellite vehicle, which then uses a separate transmitter to relay the radio signal to a receiver on another ground station, potentially located on another continent. This first important use of communications satellites for distributing television programming to major communications centers around the world was different from the later use of communications satellites to broadcast directly to individual homes equipped with small satellite dishes. Direct Broadcast Satellites did not establish regular service until the 1980s. The distribution of television broadcasts around the world was the most spectacular use of the Intelsat global system, but Intelsat also played an important role relaying international telephone calls, data, facsimile, and other forms of electronic communications.

The United States had not only assumed the lead role in organizing the Intelsat global satellite communications system, but it had also pioneered early communications satellites and the live intercontinental television broadcasts made possible by these satellites. Before the introduction of communications satellites, television film footage had to be transported by plane across the major oceans. Existing undersea telephone cables did not have the capacity to transmit television broadcasts. AT&T's *Telstar 1* satellite made the first live transatlantic television broadcast between the United States and France and the United Kingdom in early July 1962. RCA's *Relay 1* satellite transmitted live television broadcasts across the Pacific Ocean for the first time, from the United

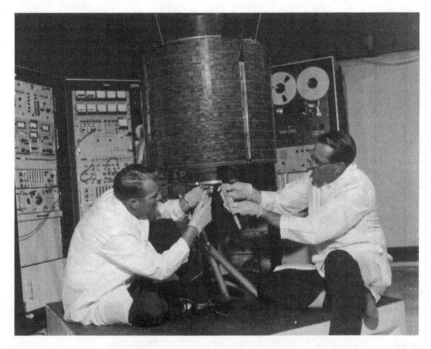

Engineers examine the *Intelsat I* communication satellite (also known as *Early Bird*), launched by NASA on April 6, 1965. It was positioned over the Atlantic Ocean to provide electronic communication, including telephone calls and television, between Europe and North America. NASA on The Commons, image 66-H-150, 1965.

Jamesburg Earth Station in California, showing satellite dish and support building. Built in 1968, it served Intelsat geosynchronous satellites parked over the Pacific Ocean. The station played a key role in the distribution of the television broadcast of the Apollo 11 moon landing. Photograph by Andrew Coile, 2020.

States to Japan, on November 11, 1963. *Relay 1* also broadcast the first major live news event to an international audience, President John F. Kennedy's funeral on November 25, 1963, which was viewed across Western Europe as well as in the Soviet Union, Eastern Europe, and North Africa.[4] Both *Relay I* and Hughes Aircraft Corporation's satellite *Syncom III* relayed live broadcasts across the Pacific of the opening ceremony of the 1964 Tokyo Olympics on October 20.[5]

Three years later, an ambitious "satellite spectacular" television program called *Our World* used the new Intelsat system to demonstrate the potential of communications satellites, including fostering global understanding and unifying the peoples of the world. Fourteen nations on five continents contributed programming to the live two-hour program. Ten other countries, although they did not contribute content, participated by receiving the programming. According to the official press release from the producers, the event used "the magic of space-age electronics to flash sound and visual images across lands, seas and time zones, fusing 'yesterday', 'today' and 'tomorrow' into a globe-encircling 'now'."[6] Canada's contribution included a discussion by the popular media analyst Marshall McLuhan, who had articulated his concept of a "global village" partly in response to the potential he perceived in communications satellites.[7] The British contribution included the Beatles singing for the first time publicly their song "All You Need Is Love."[8]

Infrastructure, Systems, and Standards

This book analyzes the decisions, especially during the Kennedy administration, leading to the establishment of the Intelsat global satellite communications system in August 1964. Locked in competition with the Soviet Union for both military superiority and international prestige, President Kennedy overturned the Eisenhower administration's policy of treating satellite communications as simply an extension of traditionally regulated telecommunications. Instead of allowing private communications companies, most notably the dominant common carrier, AT&T (American Telephone and Telegraph), to set up separate systems that would likely primarily serve profitable communication routes to Europe or other major "developed" regions, the new administration decided to take the lead in establishing a single world system open to all countries.

The Kennedy administration convinced the US Congress to pass special legislation to facilitate the creation of the global system. The Communications Satellite Act of 1962 created a unique company, Communications Satellite Corporation (Comsat), which in turn helped set up and manage the global system.[9] Intelsat was formally established in August 1964 following protracted international negotiations led by the US government and Comsat, and the signing of an interim agreement among thirteen nations. All countries were then invited to join the consortium. By the end of the decade, more than sixty countries belonged to Intelsat, with twenty-eight members operating fifty Earth stations. The system achieved worldwide coverage in 1969, when geosynchronous satellites successfully served the Pacific, Atlantic, and Indian Ocean basins. Geosynchronous satellites are a special type of satellite that appears to an observer on Earth to remain stationary because they orbit above the equator at an altitude of 22,300 miles, allowing its angular velocity to match that of the Earth's surface.

This book contributes to a growing body of scholarship studying the history of infrastructure, the basic systems and services that allow societies to function.[10] The focus here is communications infrastructure on a global scale, but other important infrastructure projects include roads, canals, railroads, electricity systems, water distribution systems, subways, and air transport.[11]

Although central to economic integration, social development, and national security, infrastructure projects are often underappreciated because of their mundane or taken-for-granted status. As "part of the background for other kinds of work," according to Susan Leigh Star, they are often effectively invisible to users.[12] Nicole Starosielski has pointed out that, in the case of communications infrastructure, scholars have paid more attention to the visible media content and its reception by users than the invisible infrastructure facilitating its distribution.[13]

In this study, I draw on Thomas Hughes's analysis of large-scale technological systems to understand communications infrastructures. Most important, Hughes argues that these systems include not only technical aspects but also interrelated organizational, political, economic, legal, and business components.[14] A recent study funded by the National Science Foundation on infrastructure echoes this analysis: "When dealing with infrastructure, we need to look to the whole array of organizational forms, practices, and institutions which accompany, make possible, and infect the development of new technology."[15] The concept of "networks" is relevant too for understanding the first global satellite communications system, because it was linked to traditional telecommunications networks. But Intelsat also had a crucial space component; especially for this reason, as well as the additional emphasis on organizational and related social components, I primarily analyze the Intelsat project as a large-scale socio-technological system rather than as a traditional telecommunications network.

An important aspect of infrastructure projects is the development of relevant technical standards. Standards and standardization, according to JoAnne Yates and Craig Murphy, have "come to provide a critical infrastructure for the global economy."[16] Standards bind together infrastructures, including the social and institutional components. They are also taken for granted and invisible, which helps give them a sense of being inevitable or uncontroversial. But this assumed harmony masks the conflicts and complex negotiations involved in standards setting.[17] Using extensive archival sources, including from recently declassified collections, I recover the complex work involved in creating technical and organization standards, including fundamental standards connected with international radio spectrum policy as well as the initial global satellite communications infrastructure they supported.

I am particularly interested here in the international dimensions of infrastructure projects, specifically their importance for understanding the history of the United States in the world and the history of US foreign relations. "Some of the most exciting new work in U.S. foreign relations," historian Daniel Immerwahr recently argued, "is moving beyond the study of ideas and intentions towards the study of infrastructure." Examining the US role in global infrastructure projects is particularly important for understanding how the post–World War II world was constructed and how the United States "managed to install itself at the center of it."[18] This book thus focuses on the efforts of US officials, especially during the period of the Kennedy administration, to project their own vision on the rest of the world through the establishment of the first global satellite communications system.

Early History of Cable and Radio

International communications, of course, predated the Intelsat system. Rapid global communications first became a reality with the establishment of undersea telegraph cables in the mid- and late nineteenth century and with the international point-to-point radio networks organized during the early twentieth century. Cables and radio provided complementary and alternative models for the organizers of the Intelsat system. A brief overview of the history of cable and wireless in a global context, with an emphasis on the implications for the United States, is necessary to understand the precedents and patterns established during the earlier radio and cable eras.

The electric telegraph integrated markets and created new distribution networks that allowed information to move strikingly faster. Time and space were increasingly compressed, especially on a global scale. Before the first undersea telegraph cable was laid successfully across the Atlantic Ocean in 1866, information such as news needed to be physically transported by ship from North America to Europe. The cable dramatically cut this time from weeks to minutes.

Although the United States was an exception, national governments generally operated telegraph lines built on land. Private companies primarily interested in commercial considerations, however, generally built and operated the undersea cables that helped integrate different parts of the world during the nineteenth century. Private companies owned nearly 90 percent of the more than 150,000 miles of undersea cables laid during the period up to 1892.[19]

An association of British cable companies, known as the Eastern and Associated Telegraph Companies, dominated undersea cables during the late nineteenth century. By 1900, the group owned approximately one-third of the

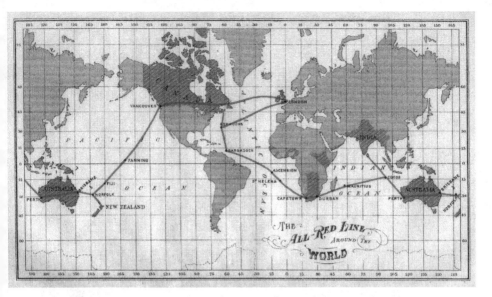

The All-Red Line of telegraph cables (colored red in the original map) connecting British-imperial territory (drawn in 1902 or 1903). Created by George Johnson.

total length of undersea cables in the world, and it had an almost complete monopoly of the routes between the United Kingdom and Central and South America, as well as complete control between Britain and India and Australia.[20]

The Eastern and Associated Companies benefited from British control of the crucial technologies and raw materials necessary for undersea cables. British plantations in Southeast Asia supplied the insulation used with undersea cables, a latex material made from the sap of the gutta percha tree. British companies led the world in cable manufacturing and in cable-laying ships with the most experienced crews. British firms also made some of the most important innovations in undersea cables during this period.[21] The dominant position of the Eastern and Associated Companies allowed the United Kingdom to remain at the center of global communications throughout the nineteenth century and into the early twentieth century.[22]

Especially by the early twentieth century, other major countries—notably France, Germany, and the United States—grew increasingly worried about British influence over the content of communications sent over cables and responded by laying their own. British censorship of news sent over cables during the Boer War helped fuel these fears.[23] Government support for undersea cables became more important during this period, partly because the national security implications also became increasingly clear. The percentage share of undersea cables owned by governments increased from approximately 10 percent to 20 percent during the "new imperialism" era, from 1880 to 1910, and state subsidies increased dramatically.[24] France, in particular, heavily subsidized new cables to link different parts of its own empire.[25] The completion by the British and key British colonies of the first undersea cable across the Pacific Ocean in 1902 (from Vancouver to New Zealand and Australia), linking the entire empire, stimulated this imperial competition for cables. The globe-circling All-Red Line of cables, which global maps traditionally colored in red, interacted only with British-imperial territory.[26]

The total length of undersea cables doubled from 1892 to 1923, but mainly at the expense of British firms. The share controlled by British firms dropped from two-thirds of the world total at the beginning of this period to one-half in 1923. US cable companies made the largest gains, increasing their share of total cable mileage from approximately 14 percent to almost 25 percent. The American company with the largest share in 1923 was Western Union Telegraph Company.[27]

An innovation during this period, point-to-point radio communication, also began to compete with undersea cables. After establishing a successful British company that used radio to send messages to ships at sea, Guglielmo Marconi introduced high-power longwave shore stations that, by the beginning of World War I, were able to send telegraphic messages between the United Kingdom and the United States during all times of the day and night. Although expensive to construct and operate, longwave shore stations were generally cheaper than cables. But during this early period, they largely complemented cables, rather than replacing them. Cables remained in demand as the need for international communications generally outstripped the overall available capacity during this time. Innovations in cable technology during the 1920s also made them more competitive compared to longwave radio. And cable had other advantages: weather could affect the quality of radio transmissions, and radio was also not as secure as cables. The British used radio as part of an integrated communications system, allowing messages to move along different routes throughout the empire as needed.[28] France and especially Germany also invested in point-to-point radio communication, but mainly because they saw it as a means to bypass British control of undersea cables.[29]

World War I demonstrated to Americans the commercial influence and national security advantages Britain received from its control of undersea cables. After the war, the United Kingdom favored maintaining control of German cables, a move opposed by US officials. The postmaster general warned in March 1919, "The world system of international electric communication has been built up in order to connect the old world commercial centres with that world business. The United States is connected on one side only. A new system should be developed with the United States as a centre."[30]

Key officials in the United States were able to convince the US government to intervene directly in international commercial communications to challenge British influence. Most important was the decision by US officials, mainly in the navy, to work to solidify the country's technological lead in radio communications after the war. Marconi's British firm was slow to adopt innovations based on the use of continuous-wave transmitters, which produced radio signals using one constant frequency. This technology more efficiently used radio spectrum, which allowed more stations to operate without causing mutual interference, and it had the ability to transmit the human voice, which meant it opened up possibilities for radiotelephony.

In 1919, officials in the navy convinced the US company General Electric (GE), a leader in manufacturing high-speed alternators that produced continuous radio waves, to block an attempt by the US subsidiary of British Marconi, American Marconi, to purchase new high-quality alternators from GE, and instead to form a new company, in exchange receiving the US Navy's radio patents. They expected that American Marconi would then be forced to sell its assets to the new firm. GE followed through with this suggestion and established a new firm, the Radio Corporation of America (RCA). Following negotiations with the British company, American Marconi did sell its assets to RCA, which expanded into international radiotelegraphy and competed effectively with cable operators. By 1923, the firm had 30 percent of the Atlantic telegraph traffic and half of the Pacific.[31]

Unlike longwave stations, the introduction of shortwave radio stations in the mid-1920s challenged the commercial viability of undersea cables. Radio engineers initially believed that the shortwave end of the radio spectrum (wavelengths 100 meters or shorter) would not be useful for sending messages over long distances. By 1925, new research conducted by the US Navy and Marconi demonstrated that shortwave transmissions could propagate over great distances after reflecting off the ionosphere. Innovations, notably in vacuum-tube technology, also helped open up shortwaves for commercial use, not only for telegraph messages but also for telephone conversations. Marconi built shortwave stations for the use of the British Post Office. The

Post Office in the United Kingdom owned and operated stations transmitting within the empire; the first circuit, connecting Britain with Canada, opened in 1926. The Marconi Company operated stations communicating with foreign countries outside the empire.[32] In the United States, RCA established transoceanic shortwave radiotelegraphy service by 1925. In 1927, AT&T introduced commercial transatlantic radiotelephony service to London, and to the continent a year later.[33]

Because shortwave radio was substantially cheaper than cables, major adjustments in the industry were inevitable. The British Post Office charged one-sixth the rate charged by cable companies to send messages to Australia using shortwave radio. In 1927, cable companies lost nearly half their business to shortwave. Fearing for the future of cable companies, the British government forced the merger of UK communications companies representing both radio and cable into a large holding company, known as Cables and Wireless, and a communications company, known as Imperial and International Communications. Regulation by the government ensured that unprofitable but strategically important cables would continue to operate.[34]

Facing similar disruptions from shortwave radio, communications companies in the United States also discussed potentially merging during this period. But the threat from RCA was not as urgent. RCA and the dominant cable companies in the United States, Western Union and the International Telephone and Telegraph Corporation (ITT), continued to discuss merger possibilities during the 1930s and 1940s; however, the Depression and World War II effectively intervened to make the need for consolidation a lower priority. Potential opposition in Congress further stifled merger talks.[35]

The Depression and World War II also delayed development of the next major innovation in international communications, ocean-spanning undersea telephone cables, introduced only a few years before the first communications satellites. AT&T and its Bell Telephone Laboratories, rather than one of the major telegraph or radio companies, provided the necessary leadership. The major technical challenge with undersea telephone cables was signal degradation caused by the surrounding sea water. A telephone cable uses a wider band of frequencies and a greater bandwidth than a telegraph cable; different frequencies are degraded selectively under sea water, resulting in an incomprehensible voice conversation. The solution was to amplify the signal at different points along the cable. Engineers from Bell Laboratories began work on this problem beginning in the 1920s, and by the early 1950s, they had perfected amplifying repeaters using small, reliable vacuum tubes improved during the war. AT&T negotiated a joint effort with the General Post Office in the United Kingdom to lay the first telephone cable across the Atlantic in 1956. Known as TAT-1, its capacity was far superior to existing telegraph cables. Its 52 telephone channels were equivalent to 459 telegraph channels, much more than the combined capacity of all the existing cables in the North Atlantic. And the message rate was a remarkable forty-eight times faster than the best prewar telegraph cable.[36]

Four themes from these early periods in the history of international communications are especially important as background for understanding the establishment of the first global satellite communications system. The first theme was the general weakening of Britain over this time and the ascent of the United States in global communications. By the 1920s, the United States had become a preeminent industrial power and was challenging Britain's control of undersea cables. US telecommunications companies first contested British communications interests in Latin America during this period.[37] US dominance in shortwave radio then finally undermined British hegemony. Britain was further weakened by World War II. And its monopoly of empire communications ended in November 1945, when it agreed to partnership arrangements with Commonwealth countries that allowed the nationalization of the local assets of Cable and Wireless. As Daniel Headrick has written, "The old communications system centered on London, which had served Great Britain well for seventy years, was disintegrating along with the empire it had served."[38]

The second theme relates to what happens when a new communications system is introduced and whether it inevitably will undermine an established industry. The history of cable and radio shows that the old and the new can both continue and complement one another. Cables were initially not threatened by longwave radio, and even after shortwave demonstrated its commercial superiority, support for telegraph cables continued. Cables maintained advantages over radio, in particular their reliability and the increased security they provided. Both wars demonstrated that messages sent over radio were vulnerable to eavesdropping.

The importance of global communications for national security points to the third major theme from the history of radio and cable—government involvement has always been important. As Daniel Headrick has argued in an overview of the history of international communications before World War II, "Private and government ownership of international telecommunications were never mutually exclusive alternatives but points along a spectrum of increasing government involvement."[39] Perhaps not surprisingly, one of the fundamental debates in the history of global communications has involved the accepted roles of the state and private enterprise. In the United States, government involvement has generally been less obvious than in other Western countries, especially Germany and France. But the US government did work to support cable companies in Latin America.[40] And although the government generally did not actively intervene in the communications industry during these earlier periods, its crucial role in the establishment of RCA shows that intervention can occur.

The fourth major theme is the growing importance of both telephony and AT&T in international communications after World War II, especially during the late 1950s and early 1960s, when the first ocean-crossing undersea telephone cables and the first communication satellites were being developed in the United States. Although a private company, AT&T was a regulated monopoly, which meant it needed to be concerned about different forms of potential government oversight. Congress had historically worried about monopoly power in communications, and these concerns had partly prevented mergers within the industry. But these earlier merger talks were among the major companies operating telegraph cables and radiotelegraph stations. AT&T was the new dominant player in the 1950s. Since it essentially held a monopoly in international telephony, the company was a potential target of antimonopoly sentiment in Congress, and this antimonopoly sentiment became especially important during debates leading to the establishment of the Intelsat satellite communications system.

Cold War Context

The crucial context for understanding the establishment of the Intelsat system during the early 1960s was the Cold War. As an infrastructure project, the global satellite communications system was made possible by developments in technology and science during that era. This study thus also contributes to the field of the history of technology and science, especially the history of Cold War technology and science. Historians who have analyzed the role of technology and science in the Cold War have tended to focus on the implications of military-related research or on the international debates about atomic energy, nuclear weapons, and new types of conventional weaponry.[41] But science and technology played a crucial role in another aspect of the Cold War: both sides in the East-West conflict attempted to use spectacular peaceful activities involving science and technology to help win the hearts and minds of average citizens in countries around the world.

Historians have used the term "global Cold War" to indicate the importance of worldwide concerns such as these, as the Cold War globalized to focus on less developed countries in Africa, Asia, and Latin America. The nuclear standoff meant that some of the most important battles between the United States and the Soviet Union involved propaganda and symbolism rather than direct armed conflict. Geopolitical leadership was determined by a country's ability to convince the world of its superior performance in advancing science and technology, especially for peaceful objectives. This book thus builds on the work of other historians who argue that the Cold War should be seen as a global struggle for public opinion and popular allegiance, especially in the Global South but also in Europe and other regions, through the use of propaganda, development tools, and other forms of "soft power"— the modern term for this program.[42] Both the United States and the Soviet Union viewed this soft-power strategy as fundamental for convincing countries to ally with them during the Cold War.

As the earliest, and arguably most important, example of the direct application of space exploration to social problems, communications satellites in general and the global Intelsat system in particular were important to the United States as a means to assert global geopolitical leadership. But this theme was also central to the entire space race. The enhancement of national prestige through technological and scientific development became especially significant with the dramatic growth in the US space program after the Soviet Union successfully orbited the first satellite, *Sputnik I*, in October 1957, and then sent the first human into space, Yuri Gagarin, in April 1961. Project Apollo, the American program to land a human on the moon initiated by President John F. Kennedy soon after the Gagarin flight, symbolized the effort to turn spectacular technological achievements into a tool of both domestic and foreign policy. Government officials sought not only to use triumphs in space to convince the world of the superiority of American political institutions but also to carry out a mandate to share with the public the knowledge acquired while building the space program.[43]

Especially during the Kennedy administration, space exploration, national security, and international communications became closely linked. The East-West Cold War conflict also became increasingly influenced by North-South tensions during these years, with the growing importance of nonaligned countries in Asia, Latin America, and Africa. The establishment of the first global satellite communications system must be understood in the context of these major developments. The use of the word "global" rather than "international" underscored not only the worldwide, or planet-encompassing, nature of the system, especially as viewed from outer space, but also the importance of establishing a system that would potentially benefit developing countries from the Global South. The earlier use of the word "international" reflected the traditional dominance of communication connections among Western countries, primarily in the Northern Hemisphere, and the traditional dominant East-West tension during the Cold War. This study of the origins of the first global satellite communications system thus sheds important new light on the transition of the Cold War from a conflict based on East-West tensions to one where North-South tensions were becoming increasingly important.

The Cold War context analyzed in this book was not only "global" but also increasingly "total." The case of satellite communications underscores the importance of a political economy of "total Cold War," in which many crucial aspects of US society became tied to imperatives of national security and geopolitical prestige.[44] When US officials negotiated the establishment of the Intelsat global system, they did not make a sharp distinction between domestic and international concerns. Most important, traditional domestic antitrust concerns were seen as closely linked to Cold War foreign policy issues, and negotiations of radio frequency standards involved a similar interplay of international geopolitical issues and US domestic matters.

With this focus on the interplay of domestic and international issues, the book contributes to a growing body of scholarship calling for a reevaluation of the nation as an analytical category in history. Thomas Bender, for example, has called for a "respatialization of historical narrative in a way that will liberate us from the enclosure of the nation." The book reinforces the transnational turn in US history by placing national experiences involving the development of satellite communications in a global context.[45]

The period of the Kennedy administration was central to the establishment of Comsat and Intelsat, and both institutions very much reflected some of the unique characteristics of the Kennedy era. In a 1984 interview conducted more than two decades after William Berman first joined Comsat, he argued that the company was "created out of the John Kennedy era of good feeling, and 'we can do anything we want to do' and 'space is our frontier' and 'you know this is Camelot' and 'if we say it won't rain during the day it won't rain during the day.'"[46] This self-confidence is also consistent with David Halberstam's account of the "best and the brightest" in the Kennedy administration. Halberstam's "whiz kids" from industry, academia, and the military also played a key role in the establishment of Comsat and Intelsat.[47]

This study thus also seeks to provide a deeper understanding of the Kennedy administration. Instead of analyzing only the involvement of elite officials and high-level policymakers, I use detailed archival records to examine the full range of actors involved in decision making leading to the establishment of the Intelsat global system, including mid- and low-level agency staff usually ignored by historians. The establishment of the global system was made possible through negotiations among several different groups and individuals with competing views about government-industry relations in both a domestic and an international context. Officials in the United States who helped establish Comsat and Intelsat debated not only the proper role of the state in technological innovation but also civil-military relations, the transfer of technology to foreign economic competitors, the relationship between developed and developing countries, and the connection between satellite communications and older, competing technologies.

The book is informed by an approach known as the social construction of technology, which is consistent with this focus on analyzing the full range of actors involved in negotiations leading to the first global satellite communications system. The approach underscores the different ways in which the development of a large-scale technological system is shaped by different groups, individuals, and organizations. It also emphasizes the need to pay attention to users. In the case of the Intelsat global system, users involved in the early negotiations included such common-carrier communications companies as AT&T and ITT, who would be among the major customers of the system.[48]

Although key decisions about the global satellite communications system were made during the Kennedy administration by officials in government and industry, earlier negotiations, especially during the 1950s, established fundamental issues and led to important early conclusions. The first chapter in the book examines this period before President Kennedy was elected in 1960. Government agencies addressed social, political, and economic issues as well as interrelated technical problems. Before communications satellites could be fully developed, different government institutions needed to develop a national policy for the new technology. Congress also began to express an interest in this process, and for the first time, key members voiced doubts about the willingness of private industry to account for geopolitical considerations important to the United States.

The development of domestic policy depended on international agreements involving frequency allocation standards. The first chapter also explores the standards negotiations conducted at a key meeting in 1959 of the International Telecommunication Union. The international delegates at this meeting for the first time agreed to assign specific frequencies for new space technologies, including communications satellites. The United States learned several important lessons at this gathering that they used later to influence global communications policy, including the need to respond more strongly to the needs of developing countries, which were becoming increasingly important during this period as the East-West Cold War conflict also exposed a growing North-South divide.

Both industry and government were interested in developing communications satellites, but AT&T conducted some of the most important early research. Long distance communications expanded dramatically during the 1950s, and AT&T was interested in alternatives to undersea cables and radio communications. AT&T developed communications satellites that orbited at a medium altitude. Since satellites at this altitude would move across the sky relative to a fixed position on the ground, Earth stations would need to track their movement. Beginning in the late 1950s, Hughes Aircraft Corporation began to work on developing geosynchronous satellites. An Earth station on the ground would not need to move to track them. The two types of satellites had different social, political, and economic implications for the future of global communications, and key government officials began to debate these interrelated issues.

During this early period, the country moved closer to what President Eisenhower called "total Cold War," especially following the Soviet success with Sputnik in 1957. National security concerns increasingly influenced domestic policy. The propaganda value of communications satellites became clearer, particularly for potentially shaping public opinion in developing countries. And although Hughes and AT&T were interested in developing commercial systems, their research was also driven by Cold War considerations. Both the military and NASA supported satellite communications research during the late 1950s.

Chapter 2 analyzes crucial decisions during the Kennedy administration that resulted in passage of the 1962 Communications Satellite Act. Although AT&T pioneered much of the early research in satellite communications, officials in the Kennedy administration initially encouraged other companies to propose alternative plans, including especially Hughes Aircraft. They doubted whether traditional arrangements for international communications developed by AT&T were appropriate in the context of the Cold War, and they feared AT&T's system was not appropriate for establishing communications service to "developing" countries. Further, officials in the Kennedy administration believed the company's commitment to new undersea cables would provide a disincentive for rapid global development of space-based technology. Concerns about AT&T's monopoly power in the communications industry also motivated critics. In the early 1960s, AT&T controlled most domestic telephone service as well as nearly all overseas telephone traffic originating in the United States, and the company was beginning to move into other forms of international communications, including data transfer and telegraph signaling. The perceived need to develop technological systems better suited to fighting a global Cold War inspired traditional critics of AT&T's monopoly power in domestic and international communications to support alternative commercial arrangements.

The focus of chapter 3 is the 1963 Extraordinary Administrative Radio Conference of the International Telecommunication Union (ITU). Known informally as the Space Radio Conference, the meeting for the first time formulated overarching international agreements for setting aside radio frequencies for all types of space activities, including satellite communications. The chapter explores the largely successful efforts of the United States to impose its vision for global communications on international spectrum policy. I first analyze the national security context that helps make sense of developments during the meeting and then examine key decisions by US officials both before and during the conference. The Space Radio Conference not only helped the United States strengthen alliances, especially with countries in Western Europe, but also provided significant lessons for nurturing relations with less developed countries. Most important, the meeting taught US officials the importance of technical assistance and technical education to poorer nations active in the ITU as well as how the planned global satellite communications system could demonstrate that all countries might benefit from space exploration.

Because Comsat was a commercial company that sold shares to the public, it needed to minimize uncertainties and potential difficulties. Not having international agreements authorizing the use of radio frequencies for satellite communications would have undermined investor confidence in the new company. Radio frequencies set aside for the exclusive use of space activities were also important to military planners, not only for military satellite communications systems but also for all aspects of rocket and missile launches. To deal with the many uncertainties in planning for the use of space frequencies, US officials proposed large blocks of frequencies at the space conference. Since the ITU did not require countries to differentiate between civilian and military users of radio frequencies, Comsat's plans were especially important in helping justify military frequency needs.

In chapter 4 I examine the decisions leading to the first interim agreement, in August 1964, establishing Intelsat, focusing on how the United States was again largely able to impose its vision for global communications on the new organization. Different organizational models were possible for the global system, but they tended to fall on a continuum between two distinct positions. A system patterned on the UN was one possibility, with decisions about governance based on a voting formula assigning each member country one vote and membership open to all countries. On the other extreme was a model generally consistent with commercial practice, in which the largest users of communications traffic would receive the greatest percentage of total available votes and would therefore have the most influence over ownership and governance.

The United States could have advocated a world organization either modeled on the UN or incorporated within the UN through its member organization the ITU. But when US officials began to consider the organization of the global satellite system the following year, they decided not to involve the United Nations and not to include all countries in initial negotiations. Since most international communications traffic operated between Western Europe and North America, US officials concluded that they should first negotiate with Western European countries to establish an initial global system based on interim agreements. The interim agreements would encourage other countries to participate, but officials decided that including non-Western countries in initial discussions would complicate planning. This decision was potentially controversial because the Kennedy administration's commitment to a global ideal had been especially motivated by a desire to involve third world countries. US officials essentially took for granted the support of Global South countries. The United States recognized that countries in Western Europe needed special attention because they would be most likely to oppose efforts to create a global system. Regionalism was central to the negotiations the Americans conducted with the Europeans, particularly with the British and the French, who threatened to establish rival systems that would especially serve their colonial or postcolonial interests. The negotiations with Americans for a global satellite communications system increasingly forced Europeans to come together as a regional block.

John Fousek's concept "nationalist globalism" helps makes sense of American involvement in the establishment of the first global satellite communications system. Traditional American nationalism inspired by a sense of exceptionalism was transformed after World War II into "nationalist globalism" by the ostensible international threat of Soviet communism.[49] The commitment, especially during the Kennedy administration, to modernization theory and development goals also helps explain United States involvement in other parts of the world. As Michael Latham has argued, this social science program provided a scientific or objective justification for US intervention in the affairs of less developed nations.[50]

Although US public officials largely worked unilaterally in deciding to establish the global Intelsat system, they did assume that once organized, the consortium would be open to multilaterally oriented decision making. At the same time, however, the Americans assumed that since the United States had developed the technology necessary for the system, including not only the satellites but also the launchers, and had the greatest need for international communications, the country would necessarily play the leading role in the new organization. US officials also justified American dominance based on their decision to organize a single global system that would allow all countries to benefit from the new space-age international communications infrastructure. In their view, the alternative they succeeded in blocking likely would have allowed AT&T to dominate the use of communications satellites for international communications, offering a service much more commercially oriented and a focus much more limited geographically. Although seemingly contradictory, nationalism and globalism could work together.

CHAPTER ONE

US Industry, the Cold War, and the Development of Satellite Communications

Communications satellites were first imagined in science fiction well before they actually existed in outer space. Historians usually credit the science-fiction writer Arthur C. Clarke, known especially for his contribution as a screenwriter for the influential 1968 film *2001: A Space Odyssey*, with giving the first fully developed account—in a 1945 article in the popular magazine *Wireless World*—of a complete satellite communications system. This chapter focuses on the period following Clarke's initial speculations and immediately before the election of President Kennedy in 1960. I analyze early decisions and developments leading to the introduction of the first communications satellites and the establishment of early government policy for the emerging communications infrastructure. Private companies, notably AT&T and Hughes, made important innovations during this period, and several government agencies debated key technical, organizational, political, legal, and economic issues. US government officials identified, for the first time, fundamental questions involving the future of satellite communications on a global scale.

National security issues linked to the Cold War increasingly influenced domestic policy, especially with the outbreak of the Korean War and after the Soviet success with Sputnik. President Eisenhower's warning about "total Cold War" reflected these developments, as domestic and foreign policy in the United States became increasingly interrelated. Most important, members of Congress and key staff members in government agencies during the Eisenhower administration began to link traditional concern about monopoly control by AT&T to new foreign policy imperatives. The Cold War increasingly became a global conflict, and spectacular innovations involving the exploration and use of outer space had not only important military implications but also crucial implications for winning over hearts and minds of peoples around the world, especially in less developed countries in Asia, Africa, and Latin America. International agreements setting aside exclusive or nearly exclusive use of radio frequencies for communications satellites, as well as space exploration in general, required the support of these countries. US officials learned important early lessons about how to impose their vision on the International Telecommunication Union from the first effort to achieve agreements for space frequencies at a key ITU meeting in 1959. Developing standards for frequency allocations would occur only through complex negotiations addressing interlinked technical and political, economic, and organizational issues. The next chapter focuses on the Kennedy administration's decision to reevaluate policy decisions made by the Eisenhower administration for satellite communications. In this chapter, I discuss early precedents for this reevaluation as government officials began to emphasize the need to move the country further toward total war and to examine the implications of domestic satellite communications policy for fighting a global Cold War.

A few individuals speculated about the use of satellites for communications at least as early as the 1920s, but Arthur C. Clarke's 1945 speculative proposal was much more detailed. He proposed that three satellites (actually manned space stations) equally spaced over the equator with angular velocity exactly matching that of the Earth's surface could theoretically provide global communications coverage (excluding regions near the poles). The special conditions allowing these "geosynchronous" satellites to appear stationary to an observer on Earth would occur at a special, relatively high orbital altitude. In this article and in another 1945 paper he circulated among the members of the British Interplanetary Society, Clarke explained the benefits of a global system for broadcasting television or other services using the ultra-high-frequency (UHF) band. Because UHF transmissions generally follow a straight-line path, stations on the Earth's surface would not be able to reach regions far beyond the optical horizon created by the planet's curvature. Since Clarke's satellites would be able to "see" nearly all points on the Earth's surface, they would have the ability to provide global television coverage.[1]

US government officials became interested in satellites immediately after World War II, inspired partly by the discovery of German plans for an ambitious program of artificial Earth satellites to accompany interplanetary travel and manned space stations. The US Navy and Army Air Forces studied the feasibility and utility of satellites for communications as well as for meteorology and mapping, making their own plans in 1946. But after being stifled by budget constraints for two years, the program was canceled completely in June 1948. These cuts reflected the dramatic demobilization that occurred at the end of the war. President Harry Truman wanted to avoid overreacting to the Soviet threat. He feared that national security concerns might ruin the economy by fueling inflation and that continuing to fund expensive military programs might transform the country into a garrison state.[2] The defense budget requested by the administration declined from approximately $80 billion in fiscal year 1945 to about $11 billion in 1948.[3] Many missile programs and high-technology weapons systems fell victim to the budget ax.

The lean years from 1945 to 1950 ended abruptly with the outbreak of the Korean War. The defense budget soared to $40 billion in fiscal year 1951 and to $60 billion during the following year.[4] With the intensification of the Cold War, the national security state expanded dramatically. A classified policy directive, NSC-68, justified the military buildup. Officials gave the highest priority to high-technology weapons, including ballistic missiles, nuclear weapons, and advanced radar defenses. Military officers, especially in the Army Air Forces, had already begun to reexamine the moribund satellite program even before the start of the war. The Army Air Forces asked the RAND Corporation, an advisory group it had established in 1945, to conduct a study of the military significance of satellites.[5]

The October 1950 RAND report provided a new justification for the use of satellites. It emphasized that satellites would provide critical reconnaissance information about Soviet defense activities and missile development, difficult if not impossible to obtain by other means. The study also underscored the psychological value of all types of Earth satellites. The first country to orbit a satellite would likely score a tremendous propaganda victory. But the report also stressed that significant legal problems remained. The legal definition of the vertical extent of a country's territory was not entirely clear. International law at the time considered the region over a country as part of a nation's territory, but a sharp boundary does not exist between the Earth's atmosphere and space. Theoretically, a country might have claimed sovereignty over parts of outer space. The report warned that the Soviets might consider a satellite orbiting over its territory as an invasion and respond aggressively. Despite these uncertainties, the government acknowledged the necessity of a reconnaissance satellite program. With plentiful funds available because of the Korean War and the implementation of NSC-68, the military gave the new project, WS-117L, a high priority. Numerous defense contractors worked on the program during the 1950s.[6]

An important suggestion made by the 1950 RAND report for finessing the legal uncertainties of satellite overflight had important implications for commercial satellite communications. The study suggested that the United States might first orbit a peaceful, scientific experimental satellite on a path that would avoid the Soviet Union. This would help set a legal precedent for freedom of space and pave the way for future military reconnaissance over Soviet territory. By the mid-1950s, events external to government planning presented an opportunity to follow through with this suggestion. An international committee of scientists asked the United States to participate in the International Geophysical Year (IGY), planned for 1957–58, by launching a scientific satellite to study global physical processes. A group of US scientists interested in upper-atmospheric and space research had first suggested this project in 1950. Lloyd Berkner, a geophysicist and an influential science administrator, played an especially important role organizing and mobilizing resources for this effort. The time frame they chose to conduct the international program not only would be a period of maximum solar activity but would also continue the tradition of the 1882 and 1932 International Polar Years, major international collaborative scientific efforts investigating the polar regions. The new project would commemorate the seventy-fifth and twenty-fifth anniversaries of the earlier events.[7]

Members of the US National Academy of Sciences successfully lobbied the federal government to support a project to orbit an experimental scientific satellite during 1957–58. Military officers and engineers who had been looking for opportunities to fulfill ambitious plans for space exploration—including Wernher von Braun, the German missile scientist and designer of the Nazis' V-2 rocket, who was now working for the American military—presented multiple proposals to the government for an IGY satellite during the early 1950s. The final alternatives included an army proposal using a military missile designed by von Braun and a plan presented by the Naval Research Laboratory (NRL) to use an improved version of a rocket originally designed by civilian contractors for high-altitude scientific research. In 1955, government officials followed the recommendation of a scientific advisory group and chose the NRL proposal. Although von Braun's rocket had a better chance of sending a satellite into orbit at the earliest possible date, the final decision also accounted for the need to minimize the military connection of the final project because of the connection to the IGY, which was organized as a peaceful activity committed to scientific investigations.[8]

The US government supported the idea of launching an IGY satellite partly to open up the legality of freedom of space, which in turn would help pave the way for the use of the reconnaissance satellites the country was secretly developing. If the Soviets launched their own satellite first, the legal problem would be eliminated. Speed, therefore, was not a major consideration. The Soviets did initiate the space age first in October 1957, with the successful orbiting of *Sputnik I*. Sputnik opened up freedom of space not only for reconnaissance satellites, which were equipped with sophisticated cameras, but also for communications satellites, similar space vehicles but equipped with sophisticated radio receivers and transmitters.[9]

In general terms, the research the federal government conducted during the years before Sputnik—mainly connected to military needs—had important relevance for satellite communications technology and ground station equipment; however, industry also took an interest in specific research and planning.[10] Long-distance communications, especially international telephony, expanded dramatically following World War II. To both supply and stimulate this demand, telecommunications companies—especially the dominant player in the United States, AT&T—sought new technological innovations that would improve the speed, quality, and capacity of this traffic. Overseas telephone and telegraph relied on undersea cables as well as long-distance radio transmission. Before 1956 radiotelephony was the only technique available for transatlantic telephone service. The first undersea telephone cable across the Atlantic (TAT-1) was installed during that year in a joint venture including AT&T, the British Post Office, and the Canadian Overseas Telecommunications Corporation. Engineers had not made many other major improvements since the laying of the last telegraph cable in 1928. And atmospheric noise seemed to place severe limitations on improvements in the quality of radio-telephony.[11]

Faced with these realities during the 1950s, AT&T and other companies were interested in pursuing alternatives to the use of cables and radio. The amount of message traffic from the United States to England rose 58 percent after the new cable was brought online, providing a clear economic incentive for telecommunications companies to pursue other innovations.[12] A potential new technique that had important implications for the development of artificial satellites used the moon as a "passive" communications satellite. Beginning as early as 1946, state governments and telecommunications companies on both sides of the Atlantic experimented with this mode of communication. Military services, especially the US Navy, were interested in new forms of secure and reliable communications techniques. Bouncing radar waves off the moon seemed superior to using shortwave transmissions on Earth because it would not depend on the unstable and unpredictable ionosphere. This form of communication also appeared less susceptible to eavesdropping, a major worry during the Cold War. The NRL pursued extensive research during the 1950s using the moon as a passive communications satellite.[13] Private companies—including International Telephone and Telegraph Corporation (ITT), which controlled a large share of international record traffic (including telegraph, data facsimile, and teletype)—experimented with this technique as an alternative to the use of undersea cables or high-frequency radio.[14]

Founded in 1920, ITT had grown rapidly during that decade to become one of the dominant companies in international communications. Using borrowed money, it purchased telephone companies throughout Latin America and connected them to long-distance lines. In 1924, AT&T, under pressure from the Justice Department to surrender its overseas interests in exchange for monopoly control in the United States, sold its foreign manufacturing subsidiaries to ITT. As a result, by 1929, ITT controlled "two-thirds of the telephones and half the cables in Latin America, one-third of the Atlantic cables, the Spanish telephones, and factories throughout the world."[15]

Officials at AT&T, in their search for new communications technologies during the 1950s, also began to consider the use of communications satellites. As early as 1954, John Robinson Pierce, director of research at AT&T's Bell Telephone Laboratories who had earned a PhD in electrical engineering from the California Institute of Technology, speculated about the commercial use of both artificial passive reflectors orbiting Earth and active satellite relays equipped with radio receivers and transmitters and thus able to receive and transmit signals. Active repeaters, he argued, would not be practical because long-lasting power sources were not available. Only passive reflectors could compete economically with cables and regular radio transmissions. In 1955, Pierce recommended the use of a reflective sphere, 100 feet in diameter, covered with aluminum foil. At an altitude of 1,000 miles, this object would be ideally suited for reflecting microwave transmissions from one ground station to another. Although it would be expensive to construct and launch into orbit, the satellite would be able to handle an enormous amount of commu-

John Robinson Pierce (1910–2002), communication satellite pioneer and director of research at AT&T Bell Telephone Laboratories. NASA on The Commons, image 52-H-001, 1960.

nications traffic, especially in comparison to the new TAT-1 cable. For this reason, according to Pierce, the initial cost could more easily be justified. His proposal thus emphasized an economic justification for the use of satellites; however, Pierce himself had broader motivations. Specifically, his interest in the commercial use of satellites was also driven by his interest in science fiction. He published his first speculation about communications satellites in 1952 in a magazine titled *Astounding Science Fiction*.[16]

Pierce's interest in satellite communications and his influential position at Bell Laboratories did not result in a full-blown program at AT&T to develop communications satellites until after the launch of *Sputnik I*, when the US government encouraged all space ventures. This fact illustrates the crucial role of the federal government in the commercial development of satellite communications. It also points to the importance of Sputnik as a catalyst for change in the United States.

Response to Sputnik

Although President Eisenhower had worried about the potential negative social and economic implications of an extreme mobilization, Sputnik shocked the country to move further toward total Cold War. The Soviet Union successfully exploited the propaganda value of Sputnik to convince the world of its superior record in supporting science and technology. Sputnik was also a psychological blow to Americans because they realized that the Soviet Union could use the same rockets that placed Sputnik in orbit to launch intercontinental nuclear missiles. Following Sputnik, the budget for high-technology weapons in the United States rose dramatically, as did the federal government's share of the research and development funds expended by all sources in the country, from 53 percent in 1955 to 65 percent in 1961. Eisenhower's recognition of the important role of science and technology to national security led to the creation of a new position in the executive branch, the special assistant to the president for science and technology, as well as a new advisory body, the President's Science Advisory Committee (PSAC). Congress passed new legislation, notably the National Defense Education Act, to help the country's schools compete with the Soviet Union. Most important, government officials reevaluated space policy and decided to increase support for two separate but linked space programs, one military and the other civilian. First, they created the Advanced Research Projects Agency (ARPA) in January 1958 within the Department of Defense (DOD). ARPA managed the entire space program until the establishment in October of the National Aeronautics and Space Administration (NASA), a civilian agency committed to scientific exploration. The new institution retained the facilities of the older National Advisory Committee for Aeronautics (NACA), which it replaced. NASA also eventually acquired important programs from the military services, including the army's rocket program in Huntsville, Alabama (led by Wernher von Braun), the army-funded Jet Propulsion Laboratory (JPL) in California, and the NRL's Vanguard Program. ARPA and the air force were responsible for military space activities and research.[17]

After Sputnik, Cold War concerns continued to stimulate military interest in experimenting with new forms of secure and reliable long-distance communications. Global competition with the Soviet Union demanded global communication. The air force was particularly interested in using satellites to facilitate communication with Strategic Air Command bombers flying over the Arctic. The first experimental communications satellites, with limited abilities, built by the US military also played an important role in demonstrating to the global community the technological superiority of the United States and, by implication, the superiority of its political, economic, and social institutions. In December 1958, President Eisenhower used ARPA's *SCORE (Signal Communication by Orbiting Relay Experiment)* satellite to relay a Christmas greeting to the world. Developed by the US Army Signal Corp and launched nearly twelve months after the successful launch of the first US satellite *Explorer 1*, which had a primary mission of measuring the radiation environment as it orbited, *SCORE* is considered the world's first communications satellite because it demonstrated for the first time that a satellite could receive signals from one location on Earth and then retransmit to a second location. The satellite could also store a message received from a ground station and then transmit the message to another ground station when directed. *Courier 1B*, the second communications satellite, which also orbited by the military, provided service in October 1960. Using more sophisticated electronics, it transmitted more than fifty million words of teletype content between a ground station in New Jersey and another in Puerto Rico during the seventeen days it functioned. Neither of these satellites was geosynchronous. Since they orbited at low altitudes, operators had to track their path as they moved across the sky relative to a fixed Earth station.[18]

Beginning as early as 1958, ARPA and the DOD organized a series of communications satellite projects, both synchronous and nonsynchronous. In 1960, the DOD decided to consolidate all work in a geosynchronous program called Advent; numerous problems, however, led the department to cancel the program two years later.[19] Defense Department officials decided they had been "overly ambitious" in trying to develop a complicated, multichannel geosynchronous satellite at a size and weight that would not exceed the capabilities of the most powerful boosters available. The engineers working on the program had particular difficulty trying to develop a positional and attitude control system that would keep the satellite fully stabilized in a precise orbital station with the antenna always pointing directly toward Earth.[20]

Despite the failure of the program, Advent did establish important precedents during this early period, including helping to entrench the view that geosynchronous communications satellites were inherently superior to other designs. Even after the DOD canceled the Advent program in 1962 and began to plan for the use of medium-altitude satellites, one of the important military officials involved in the communications satellite program, John H. Rubel, continued to argue that synchronous satellites would serve as the basis for "ultimate systems" to meet both defense and civil requirements.[21] The Advent program also shaped the research strategies of major defense contractors. General Electric (GE), ITT, and Philco Corporation received contracts to develop different aspects of the system during the Eisenhower administration. By 1960, all three companies were ready to use this research as a basis for planning commercial communications satellite operations.[22]

But two other companies, Hughes Aircraft Corporation and AT&T, conducted the most important commercial research and development during the period immediately after Sputnik. As the dominant telecommunications company in the country, AT&T had already helped develop important electronic components at its Bell Laboratories that would be useful for satellite communications, including the transistor, the maser, solar cells, and the traveling-wave tube. A traditional high-gain receiver in a ground station would add substantial noise to the transmissions relayed from a satellite. The solid-state maser could amplify the weak signal without adding significant background noise. The traveling-wave tube allowed engineers to use microwave communications over a wide range of frequencies.[23]

John Pierce had been involved in many of these developments and continued to promote satellite communications during the period leading up to Sputnik. Early in 1958, he had an opportunity to obtain government support for his passive relay proposal when he found out about upper-atmospheric research being undertaken by the NACA using a balloon similar in size to his proposed satellite. With backing from the management at AT&T, Pierce convinced NACA to adapt these experiments to the development of a passive satellite. JPL also agreed to support the project by supplying one of the Earth stations. JPL officials were mainly interested in using the experiment to help them develop new tracking and communications equipment that they could use for future space probes. T. Keith Glennan, NASA's first administrator, agreed to continue the program in 1959, after the new agency absorbed NACA. NASA launched the passive satellite *Echo I* in August 1960.[24]

Pierce and other AT&T engineers involved in the Echo project briefly considered working with ARPA (instead of with NASA) on one of its military satellites, but they considered the systems that military officials were planning, especially the geosynchronous satellites, too ambitious, complex, and expensive. They wanted to pursue plans that seemed commercially more viable, required the least amount of development, and would provide an operational system at the earliest possible time. They also did not like the idea of being "subservient" to military contractors. Before the end of 1959, the AT&T engineers worked only on passive satellites. Since NASA had agreed in November 1958 that it would only work on this type of apparatus, giving ARPA responsibility for active relays, Pierce and his colleagues probably thought it was appropriate to work with the civilian space agency.[25]

In addition to the rocket-launching and satellite apparatus, ground stations formed the third crucial part of satellite systems. They were often more expensive than the actual satellite itself. AT&T devoted most of its resources to developing this aspect of the Echo system. The ground station technology was based on developments in radar and radio astronomy. To complement JPL's Earth station in California, AT&T financed and built a second Earth station in New Jersey, incorporating the latest developments in masers and directional antenna.[26]

JPL and AT&T used *Echo I* to relay voice, telephone, facsimile, and teletype transmissions between ground stations within the United States as well as between stations in the United States and Europe. Although these experiments were impressive and demonstrated the value of communications satellites to the public, by 1960 AT&T had already begun to think about moving beyond passive relays to develop a system of active satellites that would both receive and transmit information. During the summer, the company informed the Federal Communications Commission (FCC) about plans to spend $170 million to develop a global system using fifty randomly orbiting active satellites at lower altitudes. The company contacted foreign cable partners to discuss participation in the program, assuming that the federal government would treat commercial satellite communications as simply another form of privately owned international telecommunications (like cables or radiotelephony) to be regulated by the FCC. But officials in the federal government still needed to formally establish national policy for this potential new industry. What role would NASA and other government agencies have in the development of satellite communications? Should NASA or some other government institution fund and operate satellite communications for the public or should this new development remain in private hands?[27]

During the summer and fall of 1960, Hughes Aircraft Corporation also pressured the government, especially NASA, to consider federal policy for satellite communications. Early in 1959, Hughes had decided to invest in research and planning for satellites after the air force terminated a major contract for work on a long-range interceptor needed to counter a new Soviet long-range bomber. Seeking a replacement project, the Advanced Development Laboratory at Hughes decided to investigate the potential market for geosynchronous satellite technology.[28]

AT&T chose not to use this type of satellite in its system because Pierce and other employees believed it would be too complex, heavy, and expensive. The government had not yet developed rockets capable of boosting heavy geosynchronous satellites to higher orbits. The satellites would also need more sophisticated and sensitive position and attitude controls to keep one side continuously pointed toward the Earth. Furthermore, AT&T engineers worried about time delays because of the great distances signals would have to travel (especially for two-hop links) between ground stations and the high-altitude geosynchronous satellites. Lastly, because of the small number of satellites in a geosynchronous global system, the failure of any one satellite would cause a major disruption to communications traffic.

AT&T preferred nonsynchronous, medium-altitude satellites because they required the least development. The launching and satellite technology was readily available. Operators would not need sensitive controls to keep satellites in a stationary position. But Hughes engineers pointed out that although the large number of satellites planned for AT&T's global system might provide redundancy, the launching and overall equipment costs would also be very high. Moreover, ground stations for AT&T's system would have to be much more complex. Each station would need to accurately track the satellites and smoothly switch from one satellite to another without interrupting the communications link. The Hughes engineers were aware of the military's interest in geosynchronous technology, but they believed they could build a cheaper and simpler system.[29]

By the end of 1959, Hughes engineers felt confident that they had solved some of the problems with geosynchronous satellites. They had succeeded in designing a lightweight communications satellite with reasonably simple controls for maintaining a stationary position. Hughes management gave the program its full support in spring 1960. Executives were impressed with the commercial possibilities of the proposed system, and they also believed that the company would gain excellent publicity from the venture. During 1959 and 1960, Hughes approached several potential supporters, including officials in military agencies who might provide financial assistance. Since Hughes was an aerospace company and not a common carrier like AT&T that provided communication service to customers for a fee, the company needed to sell the use of satellite channels to potential users. Executives mainly conducted negotiations with independent companies, including General Telephone and Electronics (GTE) and ITT, but also held talks with AT&T. Their primary objective in these negotiations was to get a NASA contract to build an experimental version of the satellite design. The company hoped NASA would launch the satellite and provide access to ground stations. In the meantime, Hughes was willing to continue funding development of the project with its own money. By July 1960, this amount had reached $300,000.[30]

Early Policy for Satellite Communications

Mainly in response to the efforts of AT&T and Hughes, the FCC held multiple meetings during fall 1960 to consider the future of "communications via satellite." The commission felt it should play a primary role in policy making for satellite communications based on its traditional role in licensing new uses for radio. The 1934 Communications Act directed the commission to "make available a worldwide radio communications service with adequate facilities at reasonable charge" and to "study new uses for radio . . . and generally encourage the larger and more effective use of radio in the public interest." The FCC's effort to carve out a dominant role depended partly on how it defined satellite communications. The commission argued that it was not an entirely new or radical invention but simply an extension of techniques already developed for telecommunications on the ground. The regulatory agency expected it to supplement, rather than replace, other forms of international telecommunications, including cables, radiotelephony, and radiotelegraphy.[31] The commission staff also indicated that they would "look with favor upon the establishment of that number of systems that can reasonably be supported." Had Congress and other agencies agreed to give the FCC the role it sought, the commission might have authorized competing satellite communications systems controlled by different common carriers.[32]

Toward the end of 1960, the last year of the Eisenhower administration, the activities of private companies also spurred other federal agencies to consider national policy for satellite communications. T. Keith Glennan, the director of NASA, wanted his institution to play the most important role. He felt it was NASA's "responsibility and one that we should not duck." "If some one agency doesn't step up and seek the assignment," Glennan worried, "everyone is apt to rush in and a chaotic condition can prevail."[33] One of his first actions was to arrange for termination of the agreement with the DOD that had prohibited NASA from developing active satellites.[34]

By October Glennan had established a communications satellite program within NASA and had chosen an outside contractor, Space Technology Laboratories, to manage the companies that would be involved in constructing experimental systems. During the fall, he also worked on "a statement of public policy that could be enunciated by the president" before he left office. The NASA director favored allowing private industry to "undertake the ultimate development and operation of any non-military communications satellite system." Glennan shared Eisenhower's desire to keep the United States from plunging into a race with the Soviets for spectacular space achievements of questionable scientific or commercial value. A fiscally conservative Republican, suspicious of big government, Glennan favored transferring NASA programs that had potential commercial value to private industry as soon as possible. He was particularly impressed with AT&T's proposal because the company had agreed to reimburse the government for launching and other costs.[35] Before his appointment as the first NASA administrator, Glennan had served as the president of the Case Institute of Technology in Cleveland (beginning in 1947); earlier in his career he had primarily worked as an engineer and manager in the electronics and motion picture industry.

Internal memoranda indicate that Glennan favored continuing the relationship between NASA and AT&T first established with the Echo project.[36] Glennan did not get his policy statement to the president until December 20. The president released his own public statement concerning national policy for satellite communications ten days later, but this statement essentially reiterated Glennan's view about the importance of "aggressively encouraging private enterprise in the establishment and operation of satellite relays for revenue-producing purposes."[37]

This public position, however, did not reflect the view of all government officials involved in telecommunications policy making. Attorney General William Rogers did not entirely approve of the president's policy statement of late December, mainly because it did not account for the lack of competition in international communications. He wanted the government to consider policies that would prevent AT&T from gaining monopoly control over satellite communications. Rogers believed NASA "should avoid, by all means, either creating a preferred position for AT&T or perpetuating one that already exists." He emphasized that the agency needed to "maintain every possible appearance of an 'open door' policy." In an informal conversation with Robert Nunn, special assistant to the NASA administrator, the attorney general seemed to want the agency to take a stronger stand against allowing AT&T to

Static Inflation Test of 135 Ft Satellite In Weeksville, NC
NASA Langley Research Center   6/28/1961   Image # EL-1996-00052

The satellite *Echo I*, launched by NASA in August 1960, was a balloon made of alumi-nized Mylar polyester film, which expanded after being placed in orbit. It was a pas-sive satellite that reflected radio transmissions sent from one Earth station to a sepa-rate receiving station also located on the ground. NASA on The Commons, image EL-1996-00052, 1961.

gain monopoly control over satellite communications. He felt it was "unwise" for the administration to issue a statement categorically ruling out a future role for the federal government in the "establishment and operation" of com-munications satellites for the benefit of the public; however, he also stressed that "he would not want to be coming in alone at a late hour with any imped-ing opinions."[38]

Staff members within NASA also did not entirely accept Glennan's whole-hearted support for private enterprise. Abe Silverstein, director of the Office of Space Flight Development, believed that "private industry should not have a free hand in the communications satellite business." He warned Glennan

NASA Administrator T. Keith Glennan showing Lyndon B. Johnson, at the time chairman of the Senate Aeronautical and Space Sciences Committee, the Mylar film used to construct the *Echo I* satellite. NASA on The Commons, image glennan_01.

about possible congressional opposition to any action that might allow AT&T to extend its telecommunications monopoly into space.[39]

Glennan mainly disregarded the complaints of Silverstein and other NASA officials. He complained about having to "keep hacking away at the prejudice against competitive enterprise." But Glennan did recognize that NASA had to create a competitive environment for satellite communications to forestall congressional opposition. He attempted to convince AT&T executives that they should avoid presenting the company as "a very large organization attempting to monopolize the communications field." During January 1961, in one of his last decisions as director, Glennan allowed Silverstein and Leonard Jaffe, the director of the communications satellite program, to determine policy for the first NASA-sponsored active satellite. Glennan wrote in his diary that the two men were "determined that anything short of having someone other than AT&T win the competition will be tantamount to following a 'chosen instrument' policy." When the first requests for bids went out later that month for a low-altitude active satellite, the contract eventually went to the Radio Corporation of America (RCA).[40] Even before the Kennedy administration made a formal decision rejecting Eisenhower's unqualified support for private enterprise in space, career civil servants, who would also serve the new president, were already planting the seeds of a policy reevaluation.

This reexamination also seemed necessary because the administration needed to consult more thoroughly other important groups within the government traditionally involved in telecommunications policy making. Glennan had viewed the new communications technology as mainly an application of space research. Other government bodies, including the FCC, thought it was primarily an extension of developments in telecommunications.

Reflecting the decentralized and complex nature of federal policy making, national communications policy was the responsibility of different organizations. The 1934 Communications Act established the FCC to centralize government regulation of common carriers and the use by all nongovernment stations of the radio spectrum. The act did not give the commission authority over the use of radio by the federal government. The president would direct the federal use of radio. This policy continued an earlier tradition established during the 1920s by Secretary of Commerce Herbert Hoover, who not only used his department to oversee the development of radio broadcasting but also created an advisory body, the Interdepartment Radio Advisory Committee (IRAC), to coordinate government use of radio, assign frequencies to government stations, and help prepare for international radio conferences. In 1928, President Calvin Coolidge issued an executive order effectively delegating the management of government use of radio to IRAC. The membership of IRAC included representatives from more than a dozen executive agencies. The State Department played a special role in the management of international communications. State participated directly in international radio conferences and oversaw the use by common carriers of international cables.[41]

Because the 1934 Communications Act did not authorize one government group to manage the use of the radio spectrum, the FCC and IRAC had to negotiate final decisions about assigning frequencies for users in the United States. The FCC was closely connected to Congress; the president had created IRAC. Conflicts between the two groups thus partly reflected tensions between the two main branches of government.[42]

A congressional investigation of conflicts between the FCC and IRAC in 1943 led the president to create the Telecommunications Coordinating Committee (TCC) in 1946. Officials in the executive branch wanted this group to serve as the main organization coordinating telecommunications policy in the country. But because the FCC refused to recognize its authority, the TCC never fulfilled this expectation. It functioned as an advisory body to the State Department, conducting studies of problems, collecting facts, and analyzing different policy options.[43]

The different federal agencies involved in telecommunications policy making began to consider policy for "communications via satellite" during fall 1960. Rather than provide definite answers, they mainly raised questions and problems that together underscored the fact that the new administration would have to thoroughly examine all the issues and make its own policy statement. Acting for the State Department, the TCC formed a working group to discuss satellite communications policy in September. The State Department specifically wanted to respond to AT&T's efforts to negotiate with foreign countries for use of its planned satellite system. The TCC group included six officials from the FCC, three from the air force, and one each from the navy, the Commerce Department, the army, the Federal Aviation Administration, the US Information Administration, and the State Department.[44]

During meetings in the fall, the group attempted to explore some of the major problems and questions that the government would need to engage. What would be the proper relationship between private enterprise and the government? What role would other countries have in the planned system or systems? Who in the government should negotiate agreements with these countries? Should the United States create one global system or allow multiple systems? Could private companies pay for the systems they were proposing? Would satellites support the development of international television? What role should Congress have? And should the executive branch seek new legislation? The TCC had difficulty evaluating these issues because no country or company had tested a system, and many uncertainties remained.[45]

By fall 1960 key members of Congress signaled that they wanted the government to play a more important role than simply to allow the FCC to regulate the different common carriers. A staff report, *Policy Planning for Space Telecommunications*, written at the request of Senator Lyndon Johnson's Committee on Aeronautical and Space Sciences, emphasized that policy considerations for satellite communications would involve a combination of private enterprise and national security issues. The report defined national security broadly to include "the strengths, stature, and prestige of the United States when negotiating with other nations." Significantly, it called for strong government participation in the development of a global communications satellite system that would serve Cold War aims by connecting non-Western or nonaligned countries to the United States. The federal government would need to play a central role by providing technical assistance to these countries and by making sure that they decided to link to the United States, rather than to the Soviet Union. A global satellite communications system organized by the United States would, for example, support the many African countries seeking independence and "self-realization." Instead of having to use radio or cable circuits passing through London, Paris, or other European capitals to communicate with other countries, Africans could use satellites to communicate directly with their neighbors. And the United States could support the efforts of many countries in Africa to develop television facilities by offering "transcontinental or transoceanic broadcasting of television" through satellites. In the "ideological struggle for the minds of men in these countries," the United States would need to match Soviet use of the electromagnetic spectrum for mass education, or propaganda. "Artificial satellites," according to the report, "may offer a technique by which an even more effective medium of television could be similarly employed."[46]

The Committee on Aeronautical and Space Sciences report argued that the United States should try to avoid "lending support to the image which often develops abroad and which the Soviets would try to exploit that any US program of space relays would simply enrich the private communications interests." It implied that allowing private enterprise to take the lead in satellite communications would undermine attempts by the United States to gain support for a worldwide system. Private companies would be unlikely to create unprofitable links to poor regions of the world, especially areas outside of Europe or North America. Further, the report raised concerns that AT&T's commitment to developing a global system might prove less than enthusiastic because of its existing investment in undersea cables. Congressional concern about monopoly control by AT&T must thus be understood in this wider context of international politics and foreign policy.[47]

The ideas expressed in the 1960 report of the Committee on Aeronautical and Space Sciences provided important precedents for the Kennedy administration when it reevaluated public policy for satellite communications. Historically, the report is especially significant because it not only represents one of the first examples of a general shift in global communication policy toward North-South concerns but also marks the beginning of a call for a single global satellite communications system, benefiting the entire world but also serving the Cold War interests of the United States. The report provided important precedents for the Kennedy administration when it decided to follow some of these early ideas and organize a global satellite communications system that would include poorer countries in Africa, Asia, and Latin America.

Frequency Allocation, the International Telecommunication Union, and Space Communications

One of the major policy issues that had to be dealt with before the United States could fully develop satellite communications—and launch space vehicles of all types—involved a complex interplay between economic and political factors and scientific and technological considerations. Policymakers needed to convince users of the congested electromagnetic spectrum to agree to set aside frequencies for the new technological innovation. Not only would this involve the traditional tension between IRAC and the FCC, but because of the unique international nature of satellite communications, including its potential role in Cold War politics, the United States would need to convince other countries without space capabilities to give up scarce frequencies they might use otherwise to support more accessible and immediately practical communications technologies. The history of the use of one band of frequencies, between 4 and 10 MHz, underscores the dramatic increase in demand for the finite resource: the total number of frequency listings in the world for this band went from 6,658 in 1939, to 21,456 in 1949, and to 74,284 in 1959.[48]

The International Telecommunication Union (ITU), a specialized institution within the United Nations (UN), oversees international agreements allocating different parts of the spectrum to different users. Before the Soviet Union launched the first Sputnik satellite in October 1957, the ITU's table of frequency allocations did not give special consideration to transmitters and receivers potentially used in connection with space exploration and research. *Sputnik I* transmitted on a frequency (20.005 MHz) set aside for other users of the spectrum. This practice was not necessarily a problem as long as the satellite did not interfere with other transmitters operating according to ITU policies. But because of interference with stations in the Netherlands, England, and the United States, new regulations authorizing frequencies for the new service were clearly needed.[49] Further cases of interference during the late 1950s reinforced this view.[50]

Planners recognized that they needed to reserve special radio frequencies not only for planned operational systems—such as communications, meteorological, and reconnaissance satellites—but also for individual space launches.

Experts in the United States argued that during a single launch, they would need to use as many as twenty-five different bands for tracking and sending commands to space vehicles, as well as for range safety, impact prediction, and microwave links. Each section of multistage vehicles would have its own transmitters and receivers.[51] Interference from other users could cause a rocket to veer off course, potentially threatening public safety. The first opportunity to revise the ITU's frequency table occurred in fall 1959 at the World Administrative Radio Conference.[52] The United States took the leading role in trying to convince the dozens of countries that belonged to the ITU to agree to reserve frequencies for all forms of space radio transmissions, especially satellite communications. To understand the country's involvement in the 1959 World Administrative Radio Conference and its implications for satellite communications, it is important to first understand the historical background to the ITU, especially the Cold War context.

The ITU, which dates from a meeting in Paris in 1865, is one of the earliest international, intergovernmental institutions. To coordinate separate telegraph systems across Europe, twenty nations signed two agreements at the 1865 conference, an International Telegraph Convention and the related Telegraph Regulations. An important provision of the convention called for the establishment of an International Telegraph Union, which would meet periodically to review the general rules and specific regulations. The first official meeting of the union took place in Vienna in 1868. During this meeting, the members decided to create an international bureau in a neutral location— Berne, Switzerland. The Berne Bureau would be responsible for preparing conferences and keeping track of basic information. A meeting in Berlin in 1906 established an International Radiotelegraph Convention and annexed regulations. The two administrations merged to create the International Telecommunication Union during a meeting in Madrid in 1932.[53]

The management of the radio spectrum has been one of the most important activities of the ITU. International agreements for the use of the radio spectrum have been necessary because radio waves do not simply stop at national borders. Governments have viewed the spectrum as an international common resource. Without international agreements for the management of the radio spectrum, interference could spoil its use for all users. The 1906 Radiotelegraph Conference first established a spectrum management role for the ITU. The conference made two crucial decisions—it decided to set aside discrete bands of frequencies for specific "services," and it decided that radio users had to tell the international bureau in Switzerland about their use of specific frequencies. Generally, the second decision also established an informal policy of first-come, first-served for the use of specific frequencies. Decisions about assigning specific radio services to specific parts of the spectrum became the major function of periodic administrative meetings of the ITU. This has been an ongoing issue as new technological innovations have opened up new regions of the spectrum (higher frequencies).[54]

The modern ITU originated with a crucial 1947 meeting held in Atlantic City, New Jersey. The members had agreed to join the UN two years earlier. The 1947 meeting was the first opportunity to reorganize the ITU consistent with the practices of the UN. This meant, most importantly, the establishment of an administrative council composed of eighteen members elected by the general conference of the ITU, called the Plenipotentiary Conference. The council normally met annually during the period between plenipotentiaries, carrying out the major decisions of the ITU conferences and determining the agenda for future meetings. A general secretary, responsible to the administrative council, assumed the traditional responsibilities of the old ITU bureau in Berne, Switzerland. The Switzerland connection was not completely eliminated, however. The headquarters of the new ITU was in Geneva. Plenipotentiary Conferences not only elect officials but also review and, if necessary, change the ITU's basic working guidelines, the International Telecommunication Convention.[55]

The regulatory work of the union takes place at periodic administrative conferences. Management of the radio spectrum is the responsibility of administrative radio conferences. The most important are World Administrative Radio Conferences (WARCs), which make decisions about most of the usable radio spectrum, and Extraordinary Administrative Radio Conferences (EARCs), which decide about specific blocks of frequencies.[56]

The 1947 meeting, the largest ITU gathering ever, combined a Plenipotentiary Conference and a WARC. As the conference host, major economic power, and most important user of telecommunications in the postwar period, the United States dominated the 1947 ITU Conference. The United States was especially interested in changing the traditional practice of allowing countries to effectively claim spectrum rights in a haphazard way, based especially on a policy of first-come, first-served. The old Berne Bureau in Switzerland had registered member country's claims to the use of frequencies with little oversight. Nearly all nations had abused the practice by recording fictitious uses of frequencies. The Soviet Union was particularly notorious for making excessive claims along these lines. The United States succeeded in convincing the 1947 conference to establish a new agency, the International Frequency Registration Board (IFRB), with a mandate to develop a rational allocation plan based primarily on engineering considerations and the true needs of individual countries.[57]

But these efforts to create an engineered spectrum foundered during the late 1940s and early 1950s as Cold War tensions escalated with the Korean War and other geopolitical crises. The Soviet Union viewed the IFRB as a threat to the nation's sovereign use of radio frequencies. Although the Soviets failed to eliminate the radio board, they did largely block its ability to undertake rational planning of the use of specific frequencies by different countries. The pattern that developed during the 1950s was for users to inform the IFRB after they started operations on specific frequencies; if other operators did not report interference during a two-month period, the board would automatically add the frequency use to a master list.[58]

The Soviet Union increasingly viewed the ITU as an institution largely serving the interests of the United States and its allies.[59] During the early 1950s, the Soviets primarily sought to obstruct US initiatives at ITU conferences. Stalin believed destabilizing an institution dominated by the United States was to his nation's advantage. During acrimonious discussions at the 1952 Plenipotentiary Conference in Buenos Aires, the Soviets repeatedly threatened to withhold annual contributions to the union. The Soviet Union and its allies refused to sign the final acts of the ERAC held in Geneva in 1951.[60]

The Soviet Union's treatment of the ITU partly reflected the fact that the country relied less on international communications compared to the United States and other democratic-capitalist nations. But a more important factor was Stalin's general hostility to the UN and its specialized organizations. According to historian Robert G. Weston, "Soviet policy became little more, as the Soviets put it, than preventing the United Nations from being made an instrument of imperialism and using it to check the warmongers." Stalin directed Soviet delegates to obstruct the work of the UN because of its dominance by the United States during the first postwar decade. Although the United States generally could count on support from forty-five to fifty voting members, the Soviet Union could count on the support of only about a half dozen countries at the UN. Partly because of the dominance of the United States and partly because of Stalin's indifference or hostility toward the organization, the early UN had a "disproportionate percentage of Americans on its staff." And UN technical assistance programs did not include any Soviet experts during the entire Stalin period.[61] Because of its dominant role, the

United States was much more interested than the Soviet Union in having the UN deal with international disputes. To counter UN initiatives infringing Soviet sovereignty and political interests, Soviet delegates during the Stalin era worked to undermine the authority of UN officials, most notably, Secretary General Trygve Lie, who was forced to resign late in 1952 after the Soviets boycotted and publicly insulted him.[62]

US Government Planning for the ITU during the Cold War

Officially, the State Department prepares and coordinates US participation in ITU meetings. This derives from the department's general authority to conduct foreign affairs.[63] Specifically, the department has responsibility for organizing preparatory meetings of US government agencies, selecting the US delegation, and providing delegates with general instructions. The Telecommunications Division located in the Office of Transport and Communications of the Bureau of Economic Affairs in the State Department organized many of these responsibilities during the late 1950s and early 1960s.[64]

But in practice the FCC and IRAC played key roles in decisions about US participation in international frequency allocation conferences.[65] As we have seen, the Communications Act of 1934 gave the FCC authority to assign radio frequencies to nongovernment users in the United States, while the president delegated authority to IRAC to assign frequencies to users of radio in the federal government. As a result, the two agencies have had to use complex processes of negotiation to determine not only domestic frequency allocations but also recommendations for ITU radio conferences.

To understand the context of US participation in the first ITU conferences dealing with the use of radio frequencies for outer space, it is important to first understand how US communication policy became increasingly driven by national security during the Cold War. During the first two decades after IRAC's establishment in 1922, it had been closely connected to civilian agencies, first the Commerce Department and then the FCC. A representative of the Commerce Department headed IRAC for the first eleven years. During the next eight years, the chairman of the FCC oversaw IRAC. But defense demands for radio frequencies after the outbreak of the Korean War in 1950 and the emergency drive to mobilize the country's resources to fight the Cold War resulted in a militarization of IRAC. The agency's funding became increasingly dependent on the DOD. In 1953, when President Eisenhower issued an executive order placing IRAC under the administration of the Office of Defense Mobilization (ODM), the military and national security influence became even stronger.[66] The president delegated responsibility to the director of the mobilization office to "assure maximum security to the United States in time of national emergency with a minimum interference to continuing nongovernmental requirements."[67]

This connection to mobilization agencies continued during the 1950s and early 1960s, under what was now the Office of Civil and Defense Mobilization (OCDM) beginning in 1958, and then three years later under a new successor agency, the Office of Emergency Planning.[68] An IRAC member who had represented the army served as chairman of IRAC in 1952 and 1953. Beginning in 1954 and continuing into the early 1960s, the chairman of IRAC, William E. Plummer, was a former CIA employee who also represented the mobilization agencies on the committee.[69]

Especially after 1950, the Central Intelligence Agency (CIA) increasingly linked the electromagnetic spectrum to Cold War concerns. In response to Soviet jamming of radio signals from the United States and Western Europe, the director of the CIA warned that "the Soviets are rapidly achieving the capability of launching all-out electromagnetic war against the non-Soviet world." The need for emergency preparedness and national defense demanded new institutional arrangements. The CIA recommended in 1951 that the president should establish "a central authority to coordinate and provide integrated guidance to all United States radio communications systems and organizations as necessary to prepare the United States adequately for defense against Soviet attack on our radio communication system and to mobilize the maximum United States potential for counterattack." The president responded to this recommendation not only by transferring authority for telecommunications policy in the executive branch to the ODM but also by setting up an emergency communications network called CONELRAD to link government and civilian stations.[70]

Some industry and government officials in the United States, especially members of Congress, were unhappy that no single agency decided about allocating frequencies in the United States or prepared recommendations for international conferences. As one industry critic colorfully complained in 1959, "We have two cooks and one pot."[71] The major criticism was that IRAC and the FCC arrived at final decisions through "give and take compromise" and political negotiation rather than through "adherence to logical, economic, or technical principles."[72] During the 1950s and early 1960s, critics in Congress—including Senator John O. Pastore (D-RI) and Senator Vance

Hartke (D-IN), both members of the Communications Subcommittee of the Committee on Commerce (Pastore was the chairman)—made a number of unsuccessful attempts to convince administrations to centralize communications policy, including the possibility of having a single agency that would evaluate both federal and nonfederal use of the radio spectrum.[73] But the fragmentation strengthened as national security concerns increasingly influenced IRAC decision making. In 1961, Fred C. Alexander, the telecommunications official in the mobilization agency who played a key role in decisions about frequency allocations, defended IRAC's practice of holding closed meetings, not only because open meetings would "be extremely cumbersome" but also because they would "in many cases . . . jeopardize the national defense and security."[74]

Members of Congress who favored a more centralized communication policy were especially vocal critics of IRAC's secretive proceedings. During hearings in August 1961, Senator Pastore grumbled about the inevitable outcome of his attempts to discover the reasons for IRAC decisions: "You always run into the question that you have raised about national security and there the door shuts right in your face and you can't go beyond that and we remain in the dark."[75] Testifying at the same Senate hearings, some civilian users believed that the amount of spectrum set aside for federal government agencies had risen from about 50 percent in 1951 to approximately 70 percent ten years later, "with about 40 percent exclusively Government and totally withdrawn from citizen use."[76] Fred Alexander, the telecommunications official at the mobilization agency, admitted that final decisions by IRAC and the FCC over the division of the spectrum among different users tended to favor the military because there were "certain highly classified uses" that he believed the FCC preferred "not to know and I am sure they do not know."[77] Senator Pastore was particularly critical of IRAC because he believed that the organization allowed the military services to use national security concerns to avoid proving that "proper use is being made of the bands that are allocated to the Government."[78] He thought that government decision making resulted in the defense establishment receiving "unreasonably large" blocks of frequencies that were "neither required nor used by the military services."[79] In a revealing response, Alexander admitted that he could not be sure of efficient use by all government agencies: "Not in all cases; no, sir."[80]

Because government officials serving on IRAC were users rather than regulators of the radio spectrum, they had a primary incentive to maximize requests for frequencies for their agencies, irrespective of the overall needs of the country.[81] Critics attacked the FCC for being "unwilling to counter pressure of demands of executive agencies."[82] Cold war imperatives led the commission to give in to IRAC pressure and the demands of the mobilization agencies. In 1959, the chairman of the FCC, John C. Doerfer, testified that "I think the day is past when every single person in this country is not a soldier in the next war for sure, whether it is hot or continues to be a cold one . . . We all have to recognize that. And we have to put national defense first. That's the first consideration that the FCC makes."[83]

The first opportunity to revise the ITU's frequency table came in fall 1959 at the WARC. The United States took the leading role in trying to convince the dozens of countries that belonged to the ITU to agree to reserve frequencies for space radio transmissions. This was a higher priority for the United States than the Soviet Union mainly because of the geographic differences between the two countries.[84] The Soviets only sent about a dozen delegates to the 1959 meeting; the United States, by contrast, sent more than one hundred.[85] The Soviets were less concerned about strict frequency assignments and radio interference for space exploration because they could control the radio airwaves necessary for space operations over a much larger area compared to the United States. A satellite or space vehicle would simply spend much more time over Soviet bloc territory.[86] The Americans were more dependent on other countries for tracking space vehicles. During the late 1950s, the United States established tracking stations in several countries, including Ecuador, Antigua, Chile, Peru, Australia, and South Africa.[87] During the early 1960s, agreements were extended to Spain, the United Kingdom, Nigeria, Mexico, and Zanzibar.[88] The United States needed international agreements for space frequencies especially because of the high potential of interference from a wide variety of domestic users in these different countries.[89]

In preparation for the 1959 ITU radio conference, officials connected with IRAC, the OCDM, and the FCC decided to initially request allocations only for space research. They realized that they would not likely be able to convince countries without space capabilities at this early date to agree to set aside valuable radio frequencies for advanced operational satellite systems connected with communications, navigation, and meteorology.[90] Events at the conference confirmed this fear. The United States could not even convince other nations to agree to reserve a relatively narrow range of frequencies for space research. In the face of opposition, the Americans reduced their proposal for research frequencies by 50 percent.[91]

The conference agreed to set aside thirteen narrow bands of frequencies for space communications. The United States had recommended ten of these channels. During the conference, the United Kingdom had requested one channel for space research (5250–5255 MHz), and the Soviet Union had requested two narrow bands at 40 MHz and 184 MHz. The United States was the only country to propose use of space frequencies prior to the Geneva meeting.[92] Although the US delegates requested primary status for space frequencies to minimize interference from other users, the conference agreed only to allow space researchers to share channels with other users, with no guarantee of freedom from interference. The conference also stipulated that individuals or institutions should use the space frequencies (as well as frequencies set aside for radio astronomy) only for research purposes.[93]

Only a handful of the more than eighty countries that participated in the radio conference were involved in space exploration. The United States had a difficult time convincing other countries to give up use of scarce frequencies in congested bands. Europeans were using one of the frequency channels proposed by the United States for space communications to support meteorological observations. Another band was already reserved for aeronautical radio. Although the United States could not convince the ITU to accept these requests, the other members did agree to other channel proposals, but only after they had been reduced in width and adjusted slightly to adjacent frequencies to avoid interfering with entrenched users.[94]

The policy of the new Soviet leader Nikita Khrushchev toward the UN and its specialized agencies such as the ITU differed significantly from Stalin's. Ensuring that ITU frequency agreements did not impinge on national sovereignty was still important, but in contrast to Stalin, the new leader's priority was not to work to obstruct and block proceedings as a fundamental principle. Khrushchev was willing to engage the UN. Favorable developments during the early and mid-1950s partly explain the new leader's willingness to open the country to international relations, including the steady economic development of the Soviet economy, the country's consolidation of power in Eastern Europe, and the achievement of nuclear capability. The Stalinist belief in the inevitability of wars and revolutions was replaced by a new belief in "peaceful coexistence" and a new conviction that world socialism would develop through peaceful processes. Khrushchev was willing to gamble by threatening conflict, especially by pursuing nuclear brinkmanship, but his ultimate goal was better relations with the West. By the late 1950s, this goal became increasingly important because of his desire to relieve the pressure on the consumer sector by reducing the country's defense budget through disarmament agreements.[95]

The first UN meeting that Khrushchev attended coincided with the 1959 ITU radio conference. In the speech he gave to the General Assembly in September, Khrushchev stressed the positive role of the UN in helping the world achieve peaceful coexistence and disarmament.[96] Thus, unlike during the Stalin era, Soviet opposition to United States proposals was not simply driven by a commitment to obstructing UN proceedings. The Soviets opposed United States proposals for space radio frequencies because of a general commitment to the principle of national sovereignty, but also because they were less dependent on protected radio frequencies for actual and planned space operations.[97]

Khrushchev's rise to power in the mid-1950s coincided with a period of major change in the membership of the UN and its specialized agencies. During the first decade of its existence, the membership was relatively static as the United States and the Soviet Union each refused to accept new members promoted by the other.[98] The watershed event was the UN meeting in December 1955, when sixteen states gained membership, raising the total from sixty to seventy-six. Khrushchev sought to exploit the fact that the new members added in this period mainly represented ex-European colonies with grievances against the West. During his 1959 speech to the UN General Assembly, Khrushchev demonstrated his commitment to the goals of developing countries by promoting a declaration favoring the independence of all colonial peoples.[99]

The 1959 ITU radio frequency conference was especially significant because it underscored the emergence of developing countries as a powerful new force in the UN organization. Much of the opposition to the US proposal to set aside space radio frequencies came from this segment of the ITU membership. According to the official report of the US representative at the conference, "A new element which had not arisen at earlier radio conferences was the treatment to be accorded the requirements of so called new or developing countries. Throughout the Conference reference was made frequently to these countries, and the Conference became aware at an early date of the necessity of giving earnest consideration to this rather ill-defined but very active group." Soviet opposition to US proposals to set aside special frequencies for space vehicles and space communications also demonstrated solidarity with the concerns of third world countries. For many former colonies in particular,

turning over the use of scarce radio frequencies to a few space-capable countries sounded similar to what they went through as European colonies—the
exploitation of their natural resources in a one-sided manner.[100]

An important result of the 1959 meeting was the adoption of measures to
provide technical assistance to developing countries. The ITU had been slow
to adopt the UN's general commitment to development assistance. Between
1953 and 1958, it had sent only fifty-nine experts to seventeen countries.[101]
American officials discovered that many countries in the Global South were
more interested in technical assistance than in new frequency assignments
during a meeting of a special ad hoc committee of the 1959 radio frequency
conference, created to deal with the concerns of this new group of nations:
"It was found that the needs included complete telecommunication systems,
skilled native technicians and engineers, and a knowledge of the Radio Regulations, particularly in regard to frequency assignment procedures."[102] During
the late 1950s, foreign aid and technical assistance were becoming increasingly important to both the Soviet Union and the United States as the Cold
War struggle moved out of Europe and into the third world. Eisenhower's State
of the Union address a few months after Sputnik, in January 1958, warned of
the need to fight a "total Cold War," which would especially involve a symbolic and material struggle for hearts and minds around the world, especially
in developing countries. Besides emphasizing the importance of foreign aid,
the Eisenhower administration stressed the need for psychological and political warfare to convince nations potentially attracted by Soviet achievements,
especially during the early space race, of the superiority of United States political, social, and economic institutions.[103] Foreign aid and technical assistance would also forge economic ties between the United States and developing countries, especially newly independent ex-colonies.

United States officials viewed the 1959 ITU radio conference in the context
of this symbolic and material global struggle for hearts and minds. An important lesson from the conference was the need to convince developing
countries that they would benefit from American space technology and exploration. Although US officials were only partially successful in convincing
the 1959 ITU conference to adopt its frequency proposals, they did succeed
in convincing the delegates to agree to convene an EARC in 1963 that would
specifically consider how to allocate more bands for space activities.

These concerns coming out of the UN General Assembly and the ITU radio
conference provided an important incentive for the Kennedy administration

in 1961 and 1962 to establish a global satellite communications system open to the participation of all countries in the world.[104] Kennedy was even more committed than Eisenhower to using all available resources in the global struggle, and he was also willing to go further down the path of total Cold War, which meant, most importantly, linking domestic and foreign policy considerations, including concerns about AT&T's monopoly power and the implications for building a global satellite communications system. Officials in the Kennedy administration valued communications satellites not only because of their potential military value but also because of their potential role in the Cold War battle to win hearts and minds using both spectacular and practical developments in technology and science. A global system open to all countries would go far toward convincing all countries in the world, especially developing countries, that they could benefit directly from spectacular developments in space technology.

CHAPTER TWO

The Kennedy Administration and the Communications Satellite Act of 1962

Just as the shock of Sputnik helped drive early innovations in satellite communications in the United States, the dramatic flight of the first human in outer space, Soviet cosmonaut Yuri Alekseyevich Gagarin, on April 12, 1961, led the new administration of President John F. Kennedy to reexamine government policy for communications satellites. In his famous speech to Congress on May 25 announcing plans to send a human to the moon by the end of the decade, Kennedy also announced US plans to take the lead in developing a satellite communications system that would potentially benefit the entire world.

This chapter specifically examines the decision making during the Kennedy administration leading to the passage of the 1962 Communications Satellite Act. The law established a new company, the Communications Satellite Corporation (Comsat), which had a mandate to work with other nations in setting up a global satellite communications system. Government officials debated not only the proper role of the state in technological innovation but also monopoly control, civil-military relations, the transfer of technology to foreign economic competitors, the relationship between developed and developing countries, and the connections between satellite communications and older, competing technologies. These and related issues important for government policy for satellite communications were highly contested. Because the government was internally divided, vigorous debate occurred among different agencies, within Congress, and between the administration and different groups in Congress. The resolution of these conflicts needs to be understood within the global and total Cold War context of the early 1960s that

Yuri Alekseyevich Gagarin being greeted by crowds of people in Warsaw after his first flight into outer space in April 1961. Unknown photographer, public domain, Wikimedia Commons.

moved policymakers away from making sharp distinctions between domestic and international concerns. Most important, these conflicts linked traditional domestic antitrust concerns to geopolitical goals prioritizing the establishment of a system that would potentially benefit developing countries, especially in Asia, Latin America, and Africa.

Kennedy, Outer Space, and Foreign Policy

To understand global issues involving communications during the Kennedy administration, it is important to understand Kennedy's views about the world, specifically America's relationship with other countries. Although space exploration and research expanded dramatically during his administration, Kennedy had not been particularly interested in space before he became president. One of the journalists who closely covered Kennedy's election campaign claimed, "Of all the major problems facing Kennedy when he came into office, he probably knew and understood least about space."[1] But impor-

President John F. Kennedy addressing a joint session of Congress on May 25, 1961, to announce that the "nation should commit itself to achieving the goal, before this decade is out, of landing a man on the Moon and returning him safely to the Earth." Behind him are Vice President Lyndon Johnson (*left*) and Speaker of the House Sam T. Rayburn (*right*). NASA on The Commons, image 70-H-1075, 1961.

tant campaign statements dealing with general issues of war and peace did have important implications for space policy. During the campaign, Kennedy criticized Eisenhower for allowing the country to fall behind the Soviet Union in military preparedness and world leadership. He argued that the Eisenhower administration had not responded vigorously to the Soviet challenge. Sputnik seemed to demonstrate not only the Soviet Union's general superiority in science and technology but more specifically the country's advantage in the development and deployment of ballistic missiles. This apparent "missile gap" between the two countries became a central issue in the 1960 presidential campaign.[2]

In the Senate, Kennedy had focused especially on issues related to foreign policy and international relations, particularly after 1957, when he gained a seat on the high-profile Foreign Relations Committee. As the chairman of two foreign policy subcommittees, the International Organization Affairs Subcommittee and the African Affairs Subcommittee, Kennedy stressed the importance of public diplomacy and propaganda in the symbolic battle with the Soviet Union for hearts and minds around the world. He worried especially about how Sputnik and other Soviet triumphs that occurred during the Eisenhower administration had damaged the reputation and image of the United States overseas, especially from the perspective of poorer countries in Asia, Africa, and Latin America.

But for Kennedy vigorous competition with the Soviet Union also meant that the United States needed to be open to cooperative activities. Kennedy was willing to compete with the Soviet Union, but he preferred to pursue areas of cooperation first.[3] Although Kennedy's inaugural address on January 20 emphasized his willingness to respond to the Soviet Union's challenge for world leadership, he also proposed that the two countries pursue areas of cooperation: "Together let us explore the stars, conquer the deserts, eradicate disease, tap the ocean depths and encourage the arts and commerce."[4]

Space policy was not a high priority for Kennedy during the early weeks of his presidency. He was mainly preoccupied with international crises, especially during March when he had to deal with a possible communist takeover in Laos.[5] But significantly, in his State of the Union address on January 30, he again raised the issue of cooperation with the Soviet Union in projects involving science and technology: "The United States would be willing to join with the Soviet Union and the scientists of all nations in a greater effort to make the fruits of this new knowledge available to all." Among possible areas of cooperation, he singled out the development of a "new communications satellite program."[6]

The Soviet leader Nikita Khrushchev initially tried to work with Kennedy, who entered office with a clear desire to reduce tensions, but events during spring 1961 undermined this early optimism. Khrushchev liked Kennedy's inaugural address, especially his proposal for cooperation in the peaceful uses of space.[7] Nonetheless, the Soviet Union's success on April 12 orbiting the first human around the Earth, Yuri Gagarin, and Kennedy's embarrassment over the disastrous outcome of the Bay of Pigs invasion of Cuba five days later emboldened Khrushchev to test the new American leader by challenging him globally.[8] Tensions peaked in October 1962 when Khrushchev's policy of nuclear brinkmanship led to the Cuban missile crisis.

Khrushchev's decision to intensify global tensions during the period leading to the Cuban missile crisis was thus an important context for the Kennedy administration's decision to establish a global satellite communications system open to all countries in the world, including the Soviet Union and its allies.

Especially significant was Khrushchev's decision to take back his promise to halt nuclear testing and his increased commitment to the third world. In August 1961, he accepted a KGB proposal to create "a hotbed of unrest" in Latin America by supporting local revolutionary movements.[9]

The Eisenhower administration had also tried to convince the Soviets and other countries to participate in cooperative activities involving the peaceful uses of space. In a January 1957 proposal for the international control of the testing of space vehicles, the administration had at first sought to link the use of space to disarmament or arms control initiatives.[10] This initial effort focused especially on the regulation of potential space weapons.[11] When the Soviets indicated they would try to outdo the Americans by proposing that the United Nations (UN) establish an international agency committed not only to regulating the military use of space but also to administering international cooperation in space research and exploration, the United States decided to "reassert U.S. leadership in this field and counter Soviet initiative" by taking the lead in promoting international cooperation. A report by the National Security Council in August 1958 argued, "The maintenance of our posture as the leading exponent of the use of outer space for peaceful purposes requires that the United States take in the General Assembly an imaginative and positive position."[12]

After deciding to separate military and peaceful uses of outer space, the State Department developed a proposal for the November 1958 regular session of the UN General Assembly, calling for the establishment of an ad hoc UN committee that would make recommendations for areas of international cooperation in space exploration and research under the auspices of the United Nations.[13] US officials likely decided not to deal with the use of outer space for military purposes because they did not want to open up debate about the country's highly classified reconnaissance satellite program. They also recognized the need to keep the space cooperation proposal from becoming bogged down in the intractable issue of disarmament. Agreement with the Soviets on disarmament for outer space, they realized, simply would not be possible at this time.[14] They thought it more appropriate to have a "reconstituted Disarmament Commission" consider the military uses of space.[15]

In 1959, the UN General Assembly agreed to establish a new committee called the UN Ad Hoc Committee on the Peaceful Uses of Outer Space. Because the Soviet Union had at first submitted a separate proposal "linking the peaceful uses of outer space to disarmament," the Americans viewed the Soviets' agreement to treat the two areas separately as a major victory.[16] The

Soviet Union and its allies did not block the establishment of the ad hoc committee, but they did not accept invitations to participate. The Soviets refused to cooperate "on the grounds that 'parity' of representation was not afforded between East and West."[17] They also would not agree to set up a permanent committee if it did not satisfy their concerns about parity.[18]

But in December 1959, the UN succeeded in voting for a resolution establishing a permanent Committee on the Peaceful Uses of Outer Space. They finessed the issue by simply combining the opposing proposals from the two major countries. But perhaps not surprisingly, the Soviets also refused to participate in activities of the new committee because it still did not clearly satisfy the demand for East-West parity. This meant that when Kennedy became president, the UN Committee on the Peaceful Uses of Outer Space existed on paper, but because of the lack of agreement about the membership, it still had not held any meetings.[19]

In general, Khrushchev had been desperate to achieve agreements with the Eisenhower administration during late 1959 and early 1960. Such was his optimism that he announced to the Supreme Soviet in January plans to relieve the pressure on the civilian economy by reducing the armed forces by more than one million troops.[20] But the shooting down of Francis Gary Powers's U-2 spy plane on May 1, 1960, had a chilling effect on Khrushchev's relationship with Eisenhower.[21]

Although Kennedy was not particularly concerned with the details of the American space program when he became president, he did value space exploration and research for its important role in nurturing US-Soviet cooperation. Kennedy valued space as a means to advance his foreign policy goals. He believed that agreements involving space technology and the scientific investigation of outer space would open the door to broader political and military agreements. After first deciding against a proposal from Secretary of State Dean Rusk to try to use the moribund UN Outer Space Committee to discuss outer space cooperation, key members of the Kennedy administration decided to seek bilateral agreements with the Soviets.[22] In early February, the White House established a Task Force on International Cooperation in Space to study how to proceed.[23] A draft of a report released before the Gagarin flight in early April emphasized "the U.S. preference for a cooperative rather than competitive approach to space exploration, to contribute to reduction of Cold War tensions by demonstrating the possibility of cooperative enterprise between the U.S. and the USSR in a field of major public concern."[24]

But Khrushchev did not respond favorably to Kennedy's effort to promote cooperation. Throughout the first few months of the Kennedy administration, Khrushchev continued to link space cooperation with a demand for general disarmament agreements. He was thus skeptical of Kennedy's attempts to carve out cooperative agreements in limited areas such as space technology and science.[25]

Kennedy Administration and Space Policy

Vice President Lyndon B. Johnson had played a key role in developing space policy in the Senate during the Eisenhower administration; Kennedy decided before his inauguration that he would have Johnson oversee his administration's space program. He announced in December 1960 that Johnson would take over the president's official role as the chair of the National Aeronautics and Space Council (NASC). Eisenhower generally had not used the Space Council, especially toward the end of his administration, because he preferred to directly oversee space policy rather than rely on a mediating agency.[26] He had agreed to its creation as part of a deal with majority leader Lyndon Johnson in 1958 that also involved establishing the National Aeronautics and Space Administration (NASA).

Johnson had first become interested in space as a member of the Armed Services Committee and as the chairman of the Preparedness Subcommittee. As the majority leader, he played a key role in establishing the Senate Committee on Aeronautical and Space Sciences and named himself the first chairman. He responded vigorously to Sputnik by holding twenty days of hearings investigating the country's space satellite and missile programs. Johnson saw the political advantage of taking the lead in a major high-profile national activity, but he also had a clear view of the important strategic value of space. He argued in January 1958 that "control of space means control of the world."[27]

Although some members of the Kennedy transition team favored eliminating the Space Council, it survived partly because key officials believed it would provide an important role for the vice president in the new administration.[28] Edward C. Welsh, the executive secretary of the Space Council during the Kennedy administration, argued that placing Johnson in charge of the council as the chairman was "one of the arrangements made by Kennedy" to satisfy Johnson's desire to have "some jobs to do when he became Vice President."[29] Officials in Kennedy's transition team also believed the Space Council could play an important role managing interagency conflicts by providing high-level policy direction. Although the establishment of NASA as a civilian agency was supposed to have clearly circumscribed the military's space ambitions, the air force continued to view space as its proper domain. Air force officials pressured the new administration to allow the service to expand its space activities. High-level policy direction to deal with such conflicts from an invigorated Space Council seemed necessary to the Kennedy transition team.[30] The secretary of state, the secretary of defense, the NASA administrator, and the head of the Atomic Energy Commission were the other official high-level members of the council. Kennedy told a member of Congress that the Space Council would have a "top-flight staff" and become "more than just a box on an organizational chart as it has been heretofore."[31]

Kennedy appointed Welsh executive secretary of the Space Council in March. One of his first tasks was drafting new legislation making the vice president rather than the president the head of the Space Council.[32] After earning a PhD in economics, Welsh had held a number of positions with the federal government, including most recently as legislative assistant for Missouri Democratic senator Stuart Symington. Welsh had taken a leading role during Symington's hearings on government reorganization for space in 1959 and had helped Johnson organize his Preparedness Subcommittee hearings following the launch of Sputnik. He wrote an important campaign statement for Kennedy on space, which emphasized how "control of space will be decided in the next decade. If the Soviets control space they can control earth."[33] Significantly for the development of satellite communication policy, Welsh had a reputation as a "real anti-truster."[34] Welsh generally agreed with this characterization. He not only had published in the field but also had served as chief of antitrust and cartels in Japan for more than three years after World War II.[35]

Officials in the Kennedy administration asked several individuals to oversee the work of NASA, but they all declined the invitation, probably partly because they did not want to have to deal with Johnson as the head of the Space Council. Everyone was aware of Johnson's tendency to be overbearing and domineering. Kennedy finally managed to convince James E. Webb to accept the position in February. Webb was an extremely well-connected lawyer from North Carolina who had first gone to Washington when he was elected to Congress as a Democrat. He moved to the Sperry Gyroscope Company in 1936 and became vice president seven years later. President Truman chose him to be director of the budget in 1946 and undersecretary of state from 1949 to 1952. He moved back to industry during the 1950s, serving as president of Republic Supply Company and assistant to the president of Kerr-

McGee Oil Industries. He was familiar with leading figures in the aeronautical industry through his experience as a director of McDonald Aircraft and as the general counsel of the Aeronautical Chamber of Commerce.[36]

Webb had become acquainted with the president of Kerr-McGee, Robert S. Kerr, during his tenure at the State Department. Robert Kerr was elected senator from Oklahoma in 1948 and became arguably the most powerful man in the Senate after Johnson became vice president. "When the Democrats won control of the Senate in 1960," according to Webb, "Lyndon Johnson knew he would be Vice President and worked it out with Kerr for him (Kerr) to be Chairman of [the] Aeronautics and Space Science Committee." One historian described Kerr, who was known at the time as the "uncrowned King of the Senate," as a "self-confident, overbearing, and influential senator with a thunder voice" who "walked down the Capitol's corridors" like "a 'Sherman tank in search of a target.'"[37]

In addition to powerful Washington politicians, Webb also knew many elite scientists and engineers involved in postwar government policy. He became president of the Frontiers of Science Foundation in 1955 and during the late 1950s worked part time for Educational Services, a science- and technology-oriented institution connected to MIT. Through his work with these and similar technical organizations, Webb developed working relationships with influential scientists such as Vannevar Bush, author of the crucial 1945 report to the president guiding postwar science policy *Science, the Endless Frontier*; Lloyd Berkner, the well-connected science administrator who shared Webb's interest in flying on the weekends; and Jerome Wiesner, a member of Eisenhower's Science Advisory Panel who became President Kennedy's science advisor and the president of MIT. An experienced Washington administrator and empire builder, Webb knew how to handle Johnson. Webb later reported that Kerr, who had recommended him for the position, had warned Johnson that Webb wouldn't be kicked around, "which meant Johnson never tried."[38]

Before the new administration decided specific issues connected with communications satellite policy, it began to evaluate more general space policy concerns, including the relationship between the civilian and military programs and the fate of the manned space program. A report issued in January by an ad hoc committee appointed by the president-elect, and chaired by Wiesner, argued that the administration should favor the scientific and commercial aspects of space research, including the development of communications satellites. The Wiesner Report criticized plans for manned space flight, especially the Mercury program. Although the president was sufficiently impressed with the report to appoint Wiesner as special assistant for science and technology, another report played a more important role during the first few months of the administration. The Space Science Board of the National Academy of Sciences advocated a dramatic expansion of the space program. Unlike the Wiesner study, this new report spoke out in favor of a manned mission to the moon. The strong scientific credentials of the chair of the board, Lloyd Berkner, helped give legitimacy to the study.[39]

A classified study sponsored by the air force (the Gardner Report) and completed on March 20 also recommended an expansion of the space program, but it advocated that the air force should undertake much of this work, including manned spaceflight. Even before officials finished this report, however, Congress intervened to convince the president to oppose any effort "to accentuate the military uses of space at the expense of civilian and peaceful uses." On March 9, Overton Brooks, chair of the House Committee on Science and Astronautics, warned President Kennedy that he would not "sit by and, contrary to the express intent of Congress, watch the military tail undertake to wag the space dog." Brooks believed that the Space Act establishing NASA "makes it crystal clear that the prime American mission in space is toward peaceful purposes." He warned that an overemphasis on the military side of the space program would provide Soviet propaganda with ammunition to support its portrayal of the Americans as military aggressors. He also contended that a dominant civilian program was necessary to "reap genuine economic pay-off" based on the participation of private enterprise.[40]

On March 23, Kennedy reassured Brooks that he had no intention of subordinating NASA to the Department of Defense (DOD). He emphasized that he would "rely heavily on the advice of the vice president, based on his invaluable experience with the Senate Committee on Aeronautical and Space Sciences." And he talked about reactivating the NASC under the direction of Welsh. After the administration decided to favor a strong civilian space program, NASA and the Department of Defense (DOD) tried to agree on a general division of labor. The DOD would have primary responsibility for military missiles, large solid-fueled rockets, and military communications and reconnaissance satellites. NASA would develop scientific and commercial satellites and operate the manned space program; however, the administration did promise the DOD some role in the Mercury program.[41]

NASA and the Federal Communications Commission (FCC) also tried to avoid a possible conflict when the two agencies released a memorandum of understanding on February 28 for "delineating and coordinating their respective responsibilities in the field of civil communications space activities." Both agencies agreed that statutory authority was "adequate to enable each agency to proceed expeditiously with the research and development activities necessary to achieve a commercially operable communication satellite system." But they acknowledged that special problems being studied might "result in legislative recommendations at a later date." NASA would mainly have responsibility for research and development of space technology applicable to civil communications. The FCC would mainly have authority over the "development of communications policy and the implementation and utilization of space telecommunications technology through the licensing and regulation of United States common carriers." NASA could still influence policy, however, especially through the contracts it would make with industry and through its control of launching technology.[42]

Welsh and the Space Council focused some of their first efforts on satellite communications policy because of their belief that "confusion" existed within the government concerning the new "operating entity." According to Welsh, "no interagency cooperation was set up and there was no clearly defined policy." He believed the agreements between agencies did not clearly delineate authority, and Welsh was particularly skeptical about the ability of the FCC to make policy decisions that would consider Cold War and antimonopoly interests. He feared that because FCC staff did not view satellite communications as a radical new technology, they would allow AT&T to extend its monopoly control of international communications to space.

Two sharply different understandings of the new technology of satellite communications helped drive major policy disagreements among different federal agencies. Most important, the FCC thought of communications satellites not as an exciting new space-age technology, but as an extension of traditional ground-based or undersea technologies. Asher Ende, an FCC official involved with decisions about satellite communications during this period, later admitted "to put it quite bluntly" that he and other people at the FCC viewed a satellite like a microwave tower, which "instead of being held up by steel girders, was held up by gravity." They considered it as simply "another means of providing communications." When interviewed about his experiences working with the FCC during this early period, Ende recalled that he and his FCC colleagues were dismayed to discover that people did not view communications satellites as just another "means of providing a communications path," but as a "glamorous" new space- and atomic-age technology. He claimed that before satellites, "you'd lay a cable" and "you couldn't interest people." The public did not get particularly excited, he reminisced, about the first undersea telephone cable across the Atlantic: "You'd try to tell them [about] 'This great thing that we were doing. For the first time you'll be able to talk, by voice, to Europe,' and people would say, 'Yeah?'" But, according to Ende, "mention satellites, and eyes bulge and everybody wants to get in the act."[43]

Welsh and the Space Council staff recommended that the administration increase the 1961 budget request for the satellite communications research and development (R&D) program by $10 million to experiment with different satellite technologies.[44] Kennedy accepted Welsh's recommendation during a policy meeting early on March 23, and Welsh informed Vice President Johnson about the final decision in a memorandum later in the day.[45] According to Welsh, his "relationship with the President" was such that when the Space Council met informally to formulate policy recommendations, he was the official who made "direct recommendations" to the president. The vice president normally did not get involved in specific details and specific efforts to coordinate different government agencies. Johnson expected Welsh to conduct the hard work of "getting the coordination and cooperation with the other agencies."[46] Johnson played a more important role when the NASC held formal meetings he presided over to recommend policy to the president.

Since the Eisenhower administration had assumed that a private company would contribute $10 million to NASA's program, the budget increase reflected the new administration's commitment to maintaining an active government role in the development of the new technology. Welsh believed that although the Eisenhower administration had not indicated which company would invest $10 million, "it was fairly clearly determined that it was to be the leading company in utilities." In other words, in his opinion, the administration effectively "made the arrangement that research and development in the communications satellite business was to be turned over to AT&T."[47]

Webb claimed during testimony in Congress that he participated in the decision to have the government rather than private industry contribute $10 million to satellite communications R&D. He argued, "This was such an important matter we should not start out negotiating as a government by saying that anybody who wants to negotiate with us should contribute some $10 million to the work before we would begin the negotiations."[48] When AT&T wrote NASA about their willingness to expedite satellite communications "to the maximum," Webb responded that the agency wanted "to have a good hard look at this before making commitments." He did "not want prematurely to take a stand on policy matters without a clear understanding of the ultimate results of such commitments."[49] He wanted his agency to support industry research over "a wide range of possibilities" because there was "no assurance that even as great a company as the AT&T will solve the problem in a way best calculated to serve the entire interests of the United States." Out of wide-ranging experimental research, according to Webb, "may come some new idea or some new breakthrough we do not even know about at this time."[50]

Welsh and Space Council staff also focused on satellite communications first because it was a field in which the United States was "a little bit ahead of any other country." Welsh argued, "We were ahead of the Soviets and that was the only other country we were competing with." He also believed that because it was less controversial than other programs (especially manned missions), it would be easy to gain political support. "It was a lot like being for mother-hood and an early spring," according to Welsh: "Everybody wanted better communications; in other words, it was one of these popular things."[51]

When members of the new government expressed concerns about the dominance of AT&T, they were not simply responding to the domestic issue of illegitimate economic concentration. Key individuals were also worried that AT&T might use its dominance to prematurely freeze the design of the system using a standard that might be profitable for the company but that would not help achieve national and international goals, including the rapid establishment of a truly global system serving "the new and developing countries in remote areas of Asia and Africa."[52] To achieve this goal, officials at NASA and the State Department believed that the federal government, rather than AT&T or any other international carrier, should oversee international negotiations. Traditionally, AT&T negotiated cable agreements directly with overseas government communications agencies (representing the dominant tradition in most countries of post, telephone, and telegraph agencies, known as PTTs, overseeing communications). Government foreign offices, including the State Department in the United States, traditionally played minor roles in these agreements.[53]

Satellite Communications and Commercial Interests

AT&T was not the only company interested in satellite communication R&D. Other common carriers, notably General Telephone and Electronics (GTE) and International Telephone and Telegraph (ITT), and other manufacturers, including Lockheed, Philco Corporation, and especially Hughes Aircraft Corporation, were also interested. (The geosynchronous design of the final global system chosen by Intelsat during the mid-1960s was mainly based on work done by Hughes.) The Kennedy administration supported experimentation with different types of satellites during the early 1960s. NASA declined AT&T's offer to reimburse launching expenses for its planned *Telstar 1* satellite until after the agency awarded its first contract to the Radio Corporation of America (RCA) in May 1961 for another experimental nonsynchronous communications satellite, *Relay 1*. Seven other companies—AT&T, Bendix, Collins Radio, Hughes Aircraft, ITT, and Philco—had also submitted proposals to NASA to build *Relay*.[54]

Most of these companies first took an interest in satellite communications during the late 1950s, especially after Sputnik. They were also involved in military R&D and were open to using this research for commercial innovation. By the fall of 1960, multiple companies in addition to Hughes and AT&T had conducted studies, or were in the process of conducting studies, into the commercial or civilian implications of commercial satellites, including ITT, Convair, General Electric (GE), RCA, Lockheed, and GTE.[55]

The study commissioned by Lockheed was particularly important. The Missiles and Space Division at Lockheed used Booz, Allen & Hamilton (also known as Booz Allen)—a management consulting company that increasingly specialized during the Cold War in providing advice to both government and industry about national security issues—to conduct a private business-planning study for a "Commercial Telecommunications Satellite." The letter sent with the report to Beardsley Graham, the manager of satellite planning at Lockheed, also acknowledged the assistance of Pierson, Ball & Dowd, legal consultants and specialists on government-regulated industries, and John C. Doerfer, former head of the FCC who had been pressured to resign by President Eisenhower in 1960 for vacationing on a broadcaster's yacht.[56]

The Lockheed report affirmed the commercial advantages of communications satellites relative to undersea cables. It argued that commercial international telecommunications was "of sizable proportion"; that it was growing "at a substantial rate"; and that growth in demand was "expected to exceed the capacity of existing, planned and proposed conventional communication links." In 1960, 81 percent of international telephone traffic (between continents or, to be more accurate, across large bodies of water) originated in, terminated in, or transited the United States. Between 1955 and 1959, the average annual growth rate for telephone calls placed from the United States to other places in the world was 16 percent. Telegraph messages increased annu-

Horn reflector antenna at Bell Telephone Laboratories in Holmdel, New Jersey. Built in 1959 and first used to detect radio transmissions reflected by the *Echo 1* satellite, the antenna was later adapted to receive signals from the *Telstar* satellite. NASA on The Commons, image 62-Tels-20, 1961.

ally at a smaller rate, approximately 2 to 3 percent. Although telephone calls accounted for 6 percent of message volume, they made up 23 percent of the revenue. The third major form of international message traffic, telex, started by RCA Communications in 1951, primarily as a business service using special teletypewriter equipment, had been growing at an average annual rate of 53 percent since 1951. The two major growth areas, telephone and telex, also both yielded attractive revenue—approximately twenty dollars per call. A satellite link, according to the report, "appears to be the best and most economical means of satisfying, ultimately, this growth in demand." The estimated cost of $260 million "to establish such a link" seemed justified compared to the large costs of undersea cables. Since 1956, AT&T, working with British and French carriers, had invested approximately $150 million in transoceanic cables. The British expected to spend another $250 million on a planned "globe-encircling" telephone cable. And future cables announced by AT&T were estimated to cost $165 million.[57]

By the fall of 1960, Lockheed could boast of having an "experienced satellite systems organization comprising nearly 10,000 scientists, engineers, technicians and administrative employees" that had also "developed working relationships with major subcontractors." The Booz Allen report concluded that Lockheed would need support from one of the carriers as well as potentially other manufacturers for its planned communications satellite venture. It recommended that Lockheed work toward establishing an "operating connection carrier on a 'joint venture' basis" and that "first priority should probably be given to those companies who are in both the common carrier and communications hardware businesses."[58]

Hughes had also investigated working with other companies, but Lockheed went further by seeking authorization for the joint venture from the Antitrust Division at the Justice Department. In February 1961, Beardsley Graham obtained a so-called railroad release from the Justice Department that allowed Lockheed to work with RCA and GTE in "refining the joint venture concept." The department gave a "railroad release" after an organization fully disclosed a contemplated action and the Antitrust Division concluded that it would not pursue criminal proceedings against the organization under the antitrust laws. The acting assistant attorney general in the Antitrust Division of the Department of Justice told a Lockheed representative, "We appreciate the importance and urgency of satellite communication, and we realize that the magnitude of such a project might make it difficult for a single company to carry on the necessary development work and that competitive considerations require that no one firm dominate satellite communications."[59]

The Lockheed study focused on established forms of international message traffic. It acknowledged the important potential role of satellites for transmitting or relaying television broadcasts but decided not to take this into account in its calculations because of the large uncertainties involved. Using undersea telephone cables to transmit television was generally not possible because the bandwidth requirements for television exceeded the capacity of cables. But satellites were different. Everyone recognized that communications satellites would enable, for the first time, live television from around the world. The first dramatic demonstrations during the early 1960s captured the public's imagination. *Relay 1* transmitted the first live television broadcast across the Pacific on November 23, 1963. The press gave the event major coverage, in Japan especially, because it coincided with the assassination of President Kennedy. The live news broadcasts about the assassination, which were widely viewed by the Japanese public, reinforced the spectacular implications of communications satellites for television broadcasting.[60] *Telstar 1*, an experimental satellite developed by AT&T and launched by NASA, transmitted the first live television broadcasts from an active communications satellite in July 1962. (The passive satellite *Echo 1* first relayed television across the Atlantic in 1960.) Many viewers in England and France witnessed the dramatic possibilities of live intercontinental television. They witnessed, for example, one of President Kennedy's press conferences, which *Telstar 1* relayed live across the Atlantic. And the satellite influenced popular culture when the English pop instrumental group the Tornados recorded a song titled "Telstar" during the same year. Widely played in the UK, it also reached number one in the United States on *Billboard* magazine's popularity chart.[61]

Although the Lockheed study did not attempt to predict the commercial future of international television using satellites, archival records reveal that Lockheed staff definitely speculated about different possibilities. A memorandum dated June 2, 1961, that was clearly meant as a joke and likely written by Beardsley Graham detailed an imaginary four-person official meeting on a planned "Pornosat Project" supposedly attended by Graham and another Lockheed employee as well as consultants from Booz, Allen & Hamilton and Pierson, Ball & Dowd. According to the memorandum, one of the consultants "suggested that there might be a commercial application of satellite space technology in the world-wide television circulation of hard-core pornography—photographs, printed matter and stag party flesh films." And if "sufficient programs in this category could not be found, the program schedule could be filled in with foreign art films, films condemned by the Legion of Decency and plays banned in Boston." Graham supposedly pointed out that "the State Department might have considerable interest in the Pornosat Project as one of the means of increasing our national prestige" in the art film industry, which was dominated by French, Italian, and Swedish films. And "in view of the considerable interest in the underprivileged countries and the new African nations," the participants at the meeting supposedly decided that the Booz Allen consultant at the meeting, David Hornby, "would go to Africa to make a survey of the various tribal fertility rites to determine whether they would be suitable program material for the Pornosat Project." They also decided to organize a "joint study group of other persons interested" in the subject, including officials with *Playboy* magazine "and hold a three day conference with the officials of that organization at the Playboy penthouse in Chicago." If staff from other aerospace companies joined discussions, they decided that since "such a contact might raise serious antitrust implications," they should not "rush into any such contact without first obtaining a 'railroad release' from the Justice Department."[62]

Although clearly meant as a joke, the memorandum probably was based on actual comments, perhaps after several drinks late on a Friday afternoon. The discussion underscores the dominant male culture of the aerospace industry at the time as well as the close link historically between pornography and the development of new communications technology. Especially important for this study, however, the discussion also shows that industry officials recognized the ability of communications satellites to carve out potentially radical new forms of international communications. The satirical comments about European films may also have reflected the aerospace industry's view that because it was a radical new technology, it would require new international communications policies to replace traditional ones dominated not only by AT&T but also by the European PTTs.

Kennedy, the Space Race, and Satellite Communications

Although satellite communications policy was controversial, deciding about the future of the manned lunar program that Eisenhower had vetoed in one of his last acts as president was politically much more difficult. By early April, Webb was generally convinced that astronauts could be sent to the moon and returned safely. NASA engineers had told him that it was technically feasible. But the policy question of whether it should be a priority was effectively decided by the flight of the Soviet cosmonaut Yuri Gagarin on April 12, 1961. The Gagarin flight rivaled Sputnik as a major blow to US prestige; the Americans had lost the race to orbit the first human in space. A study titled "Initial World Reaction to Soviet 'Man in Space' " by the Office of Research and Analysis of the US Information Agency (USIA) reported that media coverage was "extraordinarily heavy in almost all areas. Initial volume was comparable to that received by Sputnik I, if not greater." Comments, according to the study, "tended to acclaim the Soviet achievement generally as an epochal landmark in mankind's progress, and to move from congratulations to considering its significance in the East-West confrontation. The Soviet feat is generally held to have increased and consolidated the Soviet lead in space, and to increase Soviet military, political, and propaganda leverage."[63]

The United States was arguably winning the race to advance space science and less public technical achievements, such as guidance and computer control systems, but the Soviets had a large rocket the Americans estimated could place approximately 14,000 pounds of payload in orbit. The US one-person Mercury space capsule weighed, by contrast, only 3,900 pounds. In a letter to the vice president in late April, von Braun warned that the Soviet rocket could launch several astronauts into orbit at the same time using a capsule that might also function as a small space laboratory.[64] Even before the Gagarin flight, on March 21, Webb told the president, "We have already felt the effects of the fact that they [the Soviets] were the first to place a satellite in orbit, have intercepted the moon, photographed the back side of the moon, and have sent a large spacecraft to Venus . . . Their present position is one from which further substantial accomplishments can be expected" at a "steadily increasing pace of successful effort."[65]

President Kennedy decided not only to join the space race but to race to win.[66] Two days after the Soviet event, Kennedy called in Webb, Wiesner, and other aides to the White House to ask what could be done to regain the lead in the space race. He asked Vice President Johnson, as the head of the Space Council, to conduct an "overall survey" of American accomplishments to identify space programs that promise "dramatic results in which we could win."[67] Notes written by Hugh Dryden, deputy administrator of NASA, which he used to respond to Johnson, argued that the "lead the U.S. has taken in developing communications satellites should be exploited to the fullest. Although not as dramatic as manned flight, the direct benefits to the people throughout the world in the long term are clear." Dryden believed "U.S. national prestige will be enhanced by successful completion of this program." NASA, at the time, expected the launch of an active communications satellite in mid-1962, which he stressed would "enable live television pictures to be transmitted across the Atlantic." And Dryden's notes answering the vice president pointed out, "The continuing program will lead to the establishment of worldwide operational communications systems."[68]

In Johnson's first report to the president on April 28, he emphasized that countries "will tend to align themselves with the country which they believe will be the world leader—the winner in the long run. Dramatic accomplishments in space are being increasingly identified as a major indicator of world leadership." Among the programs identified by Johnson as fulfilling these objectives were not only the "manned exploration of the moon" but also the development of advanced satellites for worldwide communications. Satellite communications was an area in which the United States had an "advance position" that it could use to "attain world leadership" if "properly programmed with the interests of other nations."[69]

Johnson had a personal interest in broadcasting in general as well as in the specific use of television for education and international development. The head of the communications commission later recalled how Johnson had given him a long and "very impassioned talk" about the importance of communications satellites for promoting education in "underdeveloped" countries, especially through broadcasting. Johnson's specific interest in broadcasting was probably connected to his wife's involvement. She had used her inheritance to purchase and successfully operate a radio station in Austin, Texas, which then became the basis for a profitable communications company she operated. When Johnson became president after Kennedy's assassination, he supported a greater role for the country in international education, including the use of educational communications satellites.[70]

In early May 1961, Representative Overton Brooks, chair of the House Committee on Science and Astronautics, also encouraged President Kennedy to accelerate development of communications satellites, especially for the production and global distribution of television programs. "They are important," he believed, "because the nation that controls worldwide communications and television will ultimately have that nation's language become the universal tongue."[71]

Kennedy accepted with few changes Johnson's highly persuasive report, which placed an accelerated space program squarely in the context of global Cold War politics. The president's famous speech to a joint session of Congress on May 25, 1961, in which he called for sending an American to the moon by the end of the decade, also emphasized making the "most of our present leadership by accelerating the use of space satellites for world-wide communications," and linked these objectives to important domestic and international concerns, including innovative domestic programs to jump-start the economy, new foreign aid initiatives connected with the Agency for International Development, an expansion of the USIA to promote American values overseas, and new funding initiatives for civil defense.[72]

As a first step in support of worldwide satellite communications, the administration announced that it had decided to add an additional $50 million to NASA's communications satellite R&D program to experiment with different communications satellites.[73] This crucial decision allowed NASA to encourage research on systems that would provide alternatives to AT&T's. For more than a year, NASA officials had been keeping track of Hughes's development of a lightweight geosynchronous satellite, known as Syncom. When

Webb asked for the additional appropriation, he partly had in mind using the money to support Hughes's geosynchronous research.[74]

The president asked the vice president and the Space Council to coordinate different government agencies in implementing his call to support the development of space satellites for global communication. But the important agency that Welsh and others in the government thought was too close to AT&T, the FCC, had already earlier that year started to officially explore general issues involving the future of satellite communications during hearings on the allocation of radio frequencies. More important, it had opened a new major inquiry (docket 14024) in late March "into the administrative and regulatory problems relating to future authorization of commercially operated space communication systems" and had invited comments from industry and other officials by May 1.[75] Because the invited comments and final report were submitted after the Gagarin flight, they were partly influenced by this new development and Kennedy's initial response.

The twelve companies that responded to the FCC inquiry generally spoke out in favor of a communications satellite system jointly owned and operated. A number of these companies had already made extensive plans for different systems. GE, GTE (along with its subsidiary Sylvania), Bendix, Philco, and Lockheed also helped develop military satellites. The first five companies did extensive work for the military's Advent geosynchronous system, which was severely delayed by management and related problems in 1961. And government officials gave Lockheed the major contract for the WS-117L and Corona reconnaissance satellite programs. The Corona program was first proposed as a temporary solution necessary because of delays in the WS-117L program. At least three companies—RCA, GTE, and Lockheed—had already come together to undertake cooperative research on communications satellite technology.[76]

Nearly all the companies that testified at the FCC hearings reinforced the developing government position by requesting that regulators not allow any company to gain a monopoly over this new industry. They did not specifically mention AT&T by name, but AT&T was in the best position to gain control of satellite broadcasting through its dominant position as an international carrier. Although the companies that testified during the FCC inquiry in spring 1961 favored a joint venture to develop a satellite system, they disagreed about which companies would be included and how much control each company would have. AT&T and ITT argued that ownership of the system should be limited to international common carriers (like themselves). The level of ownership of each carrier would be determined by the amount of their involvement. GTE wanted to include both domestic and international carriers in the ownership of the system. Three other companies—Lockheed, GE, and Western Union—called for a broader ownership structure, not only all common carriers but also manufacturers and possibly the public. One company, Philco Corporation, proposed that after initial development of a satellite system, it would be in the best interest of the United States if the system were then turned over to be controlled and operated by "some international body, such as the United Nations."[77]

On May 24, 1961, the FCC announced tentative conclusions based on the testimony at the FCC hearings. The commission contended that "some form of joint venture by the international common carriers"—subject to certain "safeguards" and other conditions, including FCC regulatory oversight—was clearly indicated as best serving the public interest. Both "practical economics and technical limitations" indicated that only one system would be feasible. Further, the commission reasserted its belief that satellite communications would serve as a "supplement to, rather than a substitute for, existing communications systems operated by the international common carriers." "By reason of their experience and responsibility for furnishing international communications service," the international common carriers, according to the commission, "are logically the ones best qualified to determine the nature and extent of the facilities best suited to their needs and those of their foreign correspondents, with whom they have long standing and effective commercial relationships." The commission could not find a good reason to include "aerospace and communications equipment industries" in the joint venture; on the contrary, "such participation may well result in encumbering the system with complicated and costly corporate relationships, disrupting operational patterns that have been established in the international common carrier industry."[78]

In soliciting views from different companies, the FCC was particularly interested in the degree to which the limitation on the number of satellite systems would yield results "consistent with the maintenance of competition in international communications, and with the anti-trust laws and policies of the United States."[79] Because these considerations clearly involved legal issues, the commission asked the Justice Department to also respond.[80]

During the FCC hearings in May, the Justice Department expressed its belief that one jointly owned system might be consistent with the antitrust laws if all carriers and manufacturers desiring to participate were given an opportunity. The commission promised that it intended to make sure all manufacturers would have equal opportunity to supply equipment to the satellite system. It would not allow "favoritism." Further, it promised that it would not allow one company to use its control of patents for a particular standard to gain an unfair advantage over the manufacture of equipment for the satellite system, including the ground terminals. The commission also believed that it could oversee the establishment of other regulatory measures to ensure that all international common carriers (even if they did not participate through ownership) would have "equitable access to, and non-discriminatory use of, the satellite system, under fair and reasonable terms." But the commission did not decide about the "desirability or need for participation" in the joint venture by domestic common carriers. Further study and discussion with industry was needed, according to the commission, to examine all issues related to the establishment of the recommended communications satellite system. It announced plans for the organization of a committee (known as the Ad Hoc Carrier Committee) representing US international common carriers and certain relevant government agencies to explore all the issues. The commission directed the committee to finish its work by October 13.[81]

The policy group established to advise the State Department about communications, the Telecommunications Coordinating Committee (TCC), had also been studying national communications satellite policy during the first three months of Kennedy's administration. Although representatives from several executive agencies were involved in the ad hoc working group organized by the TCC to study the issue, the final report in early May generally supported many of the views expressed by the FCC. Importantly, the report also reflected the influence of the most important participant in both the TCC group and the Ad Hoc Carrier Committee, T. A. M. Craven, one of the few FCC commissioners trained as an engineer. Craven had a special interest in space communications; colleagues referred to him as the "Space Commissioner."[82] A 1913 graduate of the US Naval Academy, Craven had been involved with radio and radio policy in the government for more than forty years. The TCC report affirmed the belief that "communication via space satellite" was primarily an extension of established means of long-distance communication operated by private international common carriers, and it agreed that economic considerations and spectrum scarcity would severely limit the number of possible systems.[83]

The report did, however, raise a new issue not dealt with by the FCC. It warned that the government should not allow "the necessity for a highly reliable satellite communication system to accommodate high priority military traffic" to overshadow the development of a civil program. The report thus underscored the fact that tensions between military and civilian aspects of the space program remained. Instead of having two separate programs, the report recommended that "there appears to be no compelling reason" why the military services could not "utilize the facilities of a civil satellite system for ordinary military communication traffic under arrangements similar to those which now exists between the military and United States communication common carriers."[84]

Welsh and the Space Council

Welsh believed other government agencies besides the FCC should also be directly involved in the development of satellite communications policy. He thought the Space Council should play a central role coordinating different agencies to respond to President Kennedy's May 25 speech calling for "accelerating the use of space satellites for world-wide communications." On June 5, Welsh told the director of the Bureau of the Budget David E. Bell that he planned "to suggest to the Vice President that this be considered promptly, either through formal Council consideration or through the informal-type of procedure employed in the development of the over-all space program."[85] The council did assume a central role, but there was a delay at first when Welsh was hospitalized for a period and the vice president was away on a trip overseas. Welsh discovered to his surprise on June 5 that Webb believed he had authority to formally coordinate the "policy problem." The NASA director claimed that "he had been assigned the responsibility of coordination" by the president while the vice president was away and "that everything was proceeding satisfactorily." When the vice president returned and discovered the confusion, he met with Welsh's assistant, on June 6, and instructed him to call a council meeting for the end of the month on communications satellite systems.[86] This decision was affirmed the next day at a White House staff meeting chaired by Frederick Dutton, special assistant to the president.[87] And the president sent an official letter (drafted by Welsh) to the vice president on June 15.

Vice President Lyndon Johnson played an especially important role linking satellite communications to Cold War objectives. He directed Welsh's assistant on June 6 to seek guidance from experts in the field ("the best scientific sources that we have"), asking them specifically to "explore the ultimate in communications satellites, tell us what can be done to accelerate this program, what the by-products might be in such things as telephone conversations between remote areas in the world . . . and world wide TV, and the thrust of this should be to show us what the glories of this might be for the United States, for the free world, and to demonstrate the great difference between our system and the Soviet system."[88]

Linking developing countries to a global satellite communications system established by the United States would also help tie these countries economically to the United States through expanding global trade and commerce. US electronics and communications manufacturers could benefit through the sale of satellite Earth stations, but more important would be the indirect, long-term economic benefits resulting from the close connection between economic expansion and communications. By replacing the international cable systems operated by European colonial administrations, the global satellite communications system would potentially shift global economic patterns from Europe to the United States.

At least one company, ITT, was thinking about the implications of satellite communications for opening new markets for its equipment. In response to Vice President Johnson's survey about satellite communications, the president of the company, Harold S. Geneen, predicted a dramatic expansion of television, especially in Africa and Latin America. Communications satellites would not only be important for delivering programming to major hubs, but in twelve to fifteen years, he predicted, they would serve as "space platforms from which to beam direct TV" to individual homes. Contradicting the goals of the State Department, he favored excluding the USSR from the planned system and using the new technology "as a propaganda weapon" for advancing the "cause of democracy over communism." And he believed his company was well placed to equip the new system. The company owned telephone systems throughout Latin America and had begun to develop ground stations for communications satellites, which, he reported, would be ready by 1962. Geneen predicted a need for up to eighty new ground stations with the new system, including fifteen to twenty that would be paid for "locally by advanced countries."[89] There is little evidence that government officials were thinking specifically about economic issues such as these, which were important to the government to the extent they were connected to concerns about fighting a total Cold War. In general, Cold War national security concerns were more important than specific economic factors as motivations for government planners during this early period.

Most government officials were also not thinking about whether companies in other countries, especially Europe, would be able to contribute equipment to the planned system. During a meeting in June with the Space Council, Robert Nunn Jr., special assistant to the NASA administrator, did raise this issue briefly. He pointed out, "There was a large element of pride of prestige involved in operating and participating in an advanced communications system of the type under discussion" and he "could foresee that foreign nations would be very anxious to use their own money and contractors to build satellites to agreed specifications as part of their contribution to the operating system."[90] This would become an important matter later when the Americans conducted negotiations with other countries, especially in Western Europe. But during this initial period in the establishment of a global satellite communications system, this issue was generally not a major concern for the US government.

Using the president's June 15 letter to Johnson as a general guideline, Welsh organized a series of meetings with officials from different government agencies to review satellite communications policy. They agreed that the country needed a "high-level" policy statement from the president to replace the public statement Eisenhower had put out during the end of his term.[91] Officials from a number of agencies in the executive branch were especially critical of Eisenhower's statement because it did not address the need to accelerate the development of communications satellites as part of the space race and the need for a "global system which would serve all areas, not only those of high density traffic," thus using communications satellites as an instrument of foreign policy.[92] Nunn believed that the "lack of verbalization of national objectives governing space communications satellites was a definite shortcoming, which should be remedied quickly."[93]

During meetings discussing a draft presidential policy statement in late June and early July, Welsh invited officials from many government agencies to participate, including the FCC; however, the most influential representatives were from the main executive branch agencies officially serving on the Space Council: NASA, State Department, and DOD. Officials with these agencies believed they should be playing the most important role in determining satellite communications policy; they thought the FCC was assuming too much authority. The DOD representative "felt that there should be an agreement within the Executive Branch before the FCC went too far along the track in determining the role of private enterprise" in the planned satellite communications system.[94]

NASA officials had convinced Secretary of Defense Robert McNamara that the DOD should also support Hughes's experiments with a lightweight geosynchronous communications satellite. The agency had responsibility for the civilian space program, but it also sought to cooperate with defense space activities and research. Robert Seamans, the associate administrator of NASA, told Congress bluntly, "Everything we do will be of maximum assistance to the Department of Defense."[95] Although Secretary McNamara did not cancel the ill-fated Advent geosynchronous program until 1962, as early as March 1961 DOD officials had begun to question whether the government should continue to invest millions of dollars in the program. Webb believed the Syncom program, jointly sponsored by NASA and the DOD, would "give a great deal of experience of value to the Advent program."[96] According to John H. Rubel, who was responsible for Advent, "The Department of Defense . . . felt that if the [Syncom] experiment worked it would give us a very interesting interim communications capability."[97] NASA funded Hughes's Syncom program, and the DOD provided ground stations adapted from Advent, including air-transportable stations and one station aboard a US Navy ship.[98] The military imprint on the program was especially clear in the use of frequencies previously set aside for Advent.[99] Because of Hughes's unique contributions to geosynchronous technology, NASA awarded a "sole source" contract to the company on August 11 to build *Syncom I*.[100]

In addition to wanting to support the development of Hughes's satellite because of possible direct benefits to its own program, the DOD also viewed its support of Syncom as a way to influence the nonmilitary satellite communications system that NASA, the communications commission, and private industry were planning. The DOD wanted a separate communications satellite system it controlled for the approximately 2 percent of traffic it considered secret or classified.[101] For routine communications, the DOD and other government agencies used commercial communications networks.[102] The federal government was the largest user of the AT&T system and other national and international networks. The DOD especially, but also the State Department and other agencies, had an interest in ensuring that commercial communications systems were designed to serve their needs. Harold Brown, director of defense research and engineering, made this clear in August 1961: "Our domestic and international telecommunications systems are critical factors both in our military posture and in the Cold War struggle and, indeed, through-out the whole spectrum of conflict . . . We cannot today consider our com-munications systems solely as civil activities merely to be regulated as such, but we must consider them as essential instruments of national policy in our struggle for survival."[103] The DOD preferred a civil communications satellite system that would allow the armed forces to communicate using portable ground stations on a short notice "from those portions of the world having nonexistent or primitive communications." Because a geosynchronous system utilized relatively simple and potentially portable ground stations, the DOD believed it would do a better job achieving true global coverage.[104] This pref-erence potentially placed DOD officials in conflict with the FCC, which they feared might allow AT&T to use its industry dominance to establish a non-geosynchronous international system.

Other executive branch agencies, especially NASA and the State Depart-ment, also worried about FCC decision making. During the Space Council meetings discussing a presidential policy statement for satellite communica-tions, the State Department official was concerned "lest the current FCC ac-tion result in the approval of a system which although it might be available at an early date might not be the best system in the long run." He also worried that the FCC and other agencies were not considering the international im-plications of decisions: "Piece-meal statements by U.S. officials were present-ing a confusing pattern abroad and the Department was finding it difficult to talk to other Governments concerning our program."[105] And officials in the executive branch continued to fear that the actions being taken by the FCC would reinforce AT&T's economic dominance. A staff member from Wies-ner's Office of the Science Advisor in the White House, argued that "while no one envisioned a 100-nation organization managing such a system, neither would anyone expect the U.S. to give AT&T a blank check and tell them to go ahead and develop an international system."[106]

The tension relating to the responsibilities of the FCC, a "quasi-judicial," independent regulatory commission, and cabinet-level agencies was an im-portant context that helps explain some of the strong views in the govern-ment about the future of satellite communications. While preparing to testify before congressional committees "on the communications satellite situation," Welsh anticipated questions about the relationship between the FCC and ex-ecutive branch policymaking. He worried that members of Congress might think the White House was interfering in the work of the FCC. Whether members of Congress questioned him about this is not clear, but his prepared response emphasized that the president was not attempting "to instruct the FCC as to the specifics which fall within its competence of quasi-judicial independence." He believed the president did have "specific powers that affect independent agencies," including budgets and appointment of members. And more broadly, the president "has responsibilities for national security, foreign policy, Government allocation and use of frequencies, administration of antitrust and other legislation."[107]

Although Kennedy's June 15 letter mainly gave general guidelines for the Space Council in recommending policy, it did specifically emphasize the need for a global system. It directed the council in its deliberations that "this technology be applied to serve the rapidly expanding communications needs of this and other nations on a global basis, giving particular attention to those of this hemisphere and newly developing nations throughout the world." The letter emphasized that recommendations from the Space Council should aim to utilize fully the country's public and private resources. And it also stressed as an overall goal that "public interest objectives should be given the highest priority." Importantly, it did not explicitly state how to define the "public interest."[108] Welsh attempted to fill in some of the details during a meeting in late June when he briefed the other members on the president's June 15 letter. According to Welsh, public interest "meant the advantages that would accrue to the nation as a whole in contrast to advantages which might accrue to individuals or corporations."[109]

One of the most controversial issues was public versus private ownership and management. Kennedy's letter directed that "Policy proposals should include recommendations not only as to the nature and diversity of ownership and operation of communications systems and parts thereof, but also proposed objectives."[110] During an initial planning meeting in mid-June, Welsh pointed out "that no decision had been reached on whether the system would be Government-owned or privately-owned, and to what portions of the system such ownership would apply."[111] Theoretically, government ownership was still an option. But during this meeting officials generally assumed that private industry would take the lead. In a July 6 semipersonal letter to a medical doctor in Connecticut, Welsh emphasized his view that the main issue involving satellite communications policy was not private versus public ownership and operation (either one or the other) but "whether public interest or private interest is given the higher priority." Regardless of the type of ownership of the satellites, he argued, "private operation of the system seems most probable and practicable."[112]

Because officials believed government ownership went against traditions of private control in communications in the United States, they thought it would lead to insurmountable practical and legal difficulties. A study conducted for NASA in May had identified significant problems "if a satellite communications system were to be owned and operated by the government." The study prefaced its analysis by emphasizing that since "no one appears to recommend that the government should create an independent communication system in competition with existing systems for the business of the public, it is assumed that something less is involved, such as government ownership and operation of a system of satellites, integrated operationally as relays or links in the presently existing radio and cable networks." The main problem, according to the study, would be that this "would constitute a major and total departure from past national practice under which communication services are provided by private industry under public regulation." A new organization would be needed "of a type which does not now exist," and it would have to be integrated into the existing commercial system. The study pointed out that new legislation would be required "to define the government's responsibility and assign it to an existing or newly created agency, to define its relationship with the FCC and the State Department, and to define its relationship with existing carriers and the public." Politically, government ownership or operation of communications satellites would be difficult because the "communications industry would undoubtedly oppose any plan for government ownership." They would mobilize their considerable resources, arguing that this would reverse "long-standing" policy in the communications industry and warning that this would be a "first stop to public ownership and operation of all forms of communication."[113]

Kennedy Administration's Official Policy Statement

Welsh and the Space Council staff spent more than fifty hours in meetings during June and the first half of July drafting a policy statement for the president.[114] Welsh asked representatives of at least nine agencies and offices to comment on different drafts and then identified major concepts involved in the different submissions as well as areas of specific agreement and disagreement. The final meeting was an official Space Council meeting with major representatives of crucial institutions, notably Secretary of State Dean Rusk, Deputy Secretary of Defense Roswell L. Gilpatric, NASA Administrator James Webb, Atomic Energy Commission Chairman Glenn Seaborg, FCC Chairman Newton Minow, and Attorney General Robert Kennedy with the Justice Department. Welsh reported to the president that these official representatives on the Space Council "concurred" in two documents he submitted on July 14: a policy document containing recommendations for the development of satellite communications and a "draft public release" the president could use for the policy recommendations.[115]

To reach agreement, Welsh compromised on some issues. For example, he agreed to a request in early July from Philip J. Farley, special assistant to the secretary of state for atomic energy and outer space, asking to have a statement modified in the final public release that seemed to promise "financial assistance to developing areas." "We feel it would be unwise," Farley argued, "to issue openly an invitation for other countries to come in with their bills."[116] And he agreed to requests from the FCC to modify two significant statements. The first statement seemed to promise that satellite communications would lead to rate reductions. Better not to "go on the 'public' record," the FCC staff group wrote, "promising rate reductions." The commission advised that this would likely be a promise that "might not be fulfilled because of any number of future factors such as inflation and general cost trends." The second statement involved a policy against "excessive concentration of economic power in the communication field." The commission did not object to a general statement emphasizing that the new system would "not promote further concentration in this field." But FCC staff objected to a statement that seemed to say it was "against administration policy to have 'excessive concentration of economic power in the communication field.'" They believe it could be argued that "there is at present 'excessive concentration in the communication field' since AT&T furnishes all international Telephone services and maintains control over the facilities used or usable to furnish most other services which require a comparable bandwidth." They thought the statement could even be applied to domestic communications. "Could not the President's statement be interpreted to mean that this present concentration is against administration policy?"[117]

The documents sent to the president were recommendations; importantly, the White House decided to make some significant changes before they were publicized. One of Kennedy's special assistants, Frederick Dutton, convinced Kennedy to add antitrust statements from the Justice Department removed by the Space Council.[118] Officials in the Justice Department were especially worried that AT&T might use its manufacturing subsidiary, Western Electric, to gain monopoly control of the satellite system, even if it was jointly owned with other carriers. But, significantly, they also emphasized that antitrust decisions needed to be evaluated in the context of the "world-wide struggle" by the United States to demonstrate how the "economic system of free competitive enterprise can itself compete favorably with the Communist system of controlled monopoly." "The satellite communications system," according to the Justice Department, "can well be a prime example of the effective operation of the free enterprise system and it is, therefore, of vital importance to the national interest that no single private concern dominate satellite communication."[119]

Dutton believed strongly that "for both substantive and political reasons," the Justice Department's "anti-trust safeguards" needed to be added to the final documents. The "political reasons" involved members of Congress. He was concerned because two powerful Democratic senators—Russell B. Long of Louisiana, chair of the Subcommittee on Monopoly of the Select Committee on Small Business, and Estes Kefauver of Tennessee, chair of the Subcommittee on Antitrust and Monopoly of the Committee on the Judiciary—had recently announced that they planned to hold hearings "shortly to assure adequate anti-trust measures against AT&T in this field."[120] The White House sent copies of the communications satellite policy release to key committees in the House and the Senate immediately by messenger so that they could receive the document before it came out in the newspapers.[121]

Kennedy's public release to the press on July 24 emphasized the country's commitment to the peaceful uses of space and ensuring that all countries could benefit from space exploration. It invited all nations to "participate in a communications satellite system, in the interest of world peace and closer brotherhood among peoples throughout the world." Although it did not promise financial support to developing nations, it did state that the US government would provide "technical assistance to newly developing countries in order to help attain an effective global system as soon as practicable."[122]

Significantly, the statement seemed to rule out any form of US government ownership. It emphasized, "Private ownership and operation of the U.S. portion of the system is favored" as long as "such ownership and operation meet" specific policy requirements and safeguards that would be enforced by government agencies. Included in the requirements were the importance of rapidly constructing a system "at the earliest practicable date"; making the "system global in coverage . . . throughout the whole world as soon as technically feasible, including service where individual portions of the coverage are not profitable"; providing "opportunities for foreign participation through ownership or otherwise"; ensuring "non-discriminatory use of and equitable access to the system by present and future authorized communications carriers"; guaranteeing effective "competition, such as competitive bidding, in the acquisition of equipment used in the system"; and structuring "ownership or control which will assure maximum possible competition." The last two requirements were the "anti-trust safeguards" added by the White House. Finally, the press release emphasized that if necessary, the government might construct a separate system. This system would serve the military, but for purposes of public relations, the government did not want to state this directly.[123]

Although the statement appeared to rule out any form of government ownership or international control and management, political pressure from Congress led to a change in emphasis soon after the statement's release. On August 29, Lawrence F. O'Brien, Kennedy's former campaign manager and his special assistant for congressional relations and personnel at the White House, complimented Senator Long on the "high quality of advance preparation evidenced" in the hearings he had conducted earlier that month on the subject and assured the senator that the president's policy statement did not commit the country to a specific form of ownership. "As I understand it," he wrote, "the question of an operational commercial system is still in the research and development stage, both as to technology and as to ownership."[124]

The Kennedy administration was under increasing pressure from liberal members of Congress concerned primarily about monopoly control by AT&T. On August 24, thirty-five members of Congress, including Democratic senators Kefauver, Long, Hubert Humphrey of Minnesota, and Wayne Morse of Oregon wrote a joint letter to Kennedy informing him that they believed a decision about ownership for the planned communications satellite system needed further study and should not be concluded prematurely. They did not think the FCC had done a good job regulating AT&T in the past, and they saw no reason to trust the commission with the task of preventing AT&T from extending its control of international communications into space.[125]

The thirty-five members of Congress who wrote Kennedy urged him to avoid deciding, or allowing the FCC to decide, until he had satisfactorily dealt with the issue of preventing monopoly control. On July 25, the FCC had emphasized that it expected detailed plans about the establishment of a global satellite communications system from the Ad Hoc Carrier Committee by October 13.[126] The members of Congress thought the government should first establish an operational system before deciding about private ownership. "After such a system has become fully operational, but not until then," the liberal faction argued, "can decisions be intelligently made as to whether such a system should be publicly or privately owned and under what circumstances."[127]

Debates about Ownership

Although the State Department and the other executive branch agencies had concurred in the final recommendation of the Space Council that seemed to clearly rule against government ownership or management, senior officials in the State Department had in fact expressed strong reservations about the wisdom of having private industry play a dominant role in the planned global satellite communications system. On July 13, Harlan Cleveland, the assistant secretary of state for international affairs, told the secretary of state that he was "not entirely happy" with the draft statement the council planned to discuss at the final meeting scheduled for the next day. He asked the secretary to consider an alternative statement. Cleveland indicated that he planned to discuss this in more detail at the department meeting on the subject scheduled for 12:30 p.m. Since the secretary planned to attend the official Space Council meeting later in the afternoon, Cleveland clearly did not give him sufficient time to make major changes. He informed the secretary that the Bureau of International Organization Affairs in the State Department had approved his request, but he had not yet had an opportunity to submit it to the entire department. Cleveland's main concern was that the Space Council draft statement dealt "mainly with domestic policy considerations and gives only secondary attention to the possibility of foreign participation in ownership and management."[128]

During this period in mid-July, Cleveland was preparing a proposal for the next session of the UN General Assembly. The initiative involved a "comprehensive program for United Nations activities in peaceful uses of outer space," which contained "as one important element a statement of intention about communications satellites," including a promise that "all countries will be given an appropriate opportunity to participate in the ownership and management of the system." He was especially concerned about making sure the Space Council proposal was consistent with this General Assembly initiative. Significantly, in his forthcoming General Assembly initiative he planned to offer more details than the Space Council discussed in its recommendation, including announcing that the United States was "tentatively contemplating" an "international company with shares held by private firms, public corporations, and governments which would own the system and manage its day-today operations." And he was also seriously considering recommending that an "intergovernmental body (the ITU or a new agency)" be given the authority "to make recommendations to the international company" on such issues as the design of the system and the setting of rates.[129]

Cleveland and another officer in the Bureau of International Organization Affairs had discussed this proposal with officials from other agencies in the executive branch. Hugh Dryden, deputy administrator of NASA, thought "it would be very difficult to establish a workable international agency for outer space cooperation and questioned whether the United Nations has operational capabilities which would permit it to play such a role." Cleveland countered with the example of the "successful operation of the European Coal and Steel Community and expressed the conviction that this is the kind of approach that will be called for in the field of outer space in the next twenty years."[130]

In response to the letter from thirty-five members of Congress, Welsh deemphasized the clear language in the president's policy statement favoring private ownership. On September 18, the Space Council argued that "the policy statement provided for flexibility as to the type of ownership and operation of the U.S. portion of the system." The White House requested that the Space Council hold meetings to determine the best organizational arrangement. Instead of thinking in general terms, Welsh analyzed the situation in detail and asked officials to consider nine different options. On one extreme was private ownership restricted to the existing international carriers and based on "international use experience or upon availability of capital for such investment." The other extreme was a "government corporation, owning the U.S. satellites, and financed through appropriated funds or on earnings from annual revenues obtained from the use of the satellites." The other options between these two extremes included proposals with mechanisms to prevent a private company from dominating the consortium; those that allowed investment by other companies, including hardware manufacturers; proposals that also allowed raising capital through public stock offering but with limitations on any one company or individual; and those for special private-public partnerships such as a joint stock corporation, "with the ownership and direction divided between representatives of government and private capital."[131]

Welsh also reminded other government officials that satellite communications included multiple different components and that separate policies involving ownership and organization might be applied to different system elements. The different components included the satellite itself, the transmitting and receiving Earth stations used to communicate with satellites, ground support equipment such as tracking stations and launch site facilities, the launch vehicle for the satellite, and patents used in the satellite system that "represent a kind of property about which ownership decisions must be made." The Space Council mainly thought about different choices as "being completely U.S. owned." But Welsh recognized that it was "also possible to consider a variety of forms with foreign participation whether these are private or public in nature." He pointed out that international holdings "in private corporations, and corporate combinations using foreign subsidiaries are quite common." And examples of intergovernmental corporations also existed, for instance, the Scandinavian Airlines System.[132]

The Space Council continued to hold meetings in September, but the White House was also assuming a more active role. In late September, Dutton in the White House asked representatives participating in the meetings to submit each agency's view about the most appropriate organizational arrangement for the planned satellite communications system. They planned to discuss the submissions at a meeting on October 3 in the second-floor conference room in the White House.[133]

The different agencies officially represented on the Space Council disagreed about the form of ownership of the US aspect of the system. The Department of Justice, the State Department, and Wiesner's office all favored "some form of government ownership."[134] NASA and to a lesser degree the DOD favored private ownership. NASA did not believe organizational decisions needed to be made at the time, but the agency generally supported the FCC's decision to allow the carriers to decide the future of satellite communications. The DOD mainly opposed joint or government ownership because they thought these alternatives would require new legislation that would delay the system. The State Department believed an "international joint undertaking should own and operate the satellites and directly associated ground control facilities for the system." The department favored a flexible organization that would allow "for ownership in participating countries to be either by the Government or private interests." State Department representatives argued strongly for "Government participation or ownership of the U.S. portion." The Justice Department supported a "special statute authorizing combined Government and private ownership." If the proposed corporation ended up covering only satellites, then the department believed it should be nonprofit. Justice also favored having the US portion of the system include not only all carriers but also ideally "equipment makers." The department suggested that there "could be a corporation to manage the system funded by industry and government with the board of directors controlled by the government." Justice proposed the creation of a "government Communications Authority to own and operate the satellites only."[135]

Welsh also asked the five main staff members with the Space Council for their personal preferences regarding the organization of the planned satellite system. Four staff members favored either a government corporation or some type of organization with "joint government-private ownership." Russell Hale, Welsh's main assistant who had previously worked for the State Department, believed a government corporation could eventually be "converted to a joint private-government ownership" after the system had been "stabilized." The staff member who suggested private ownership and who believed the FCC should make the choice did not oppose the other alternatives listed by Welsh. He argued that "none of the alternatives" listed were "unacceptable."[136]

After a second meeting with the five government agencies (NASA, Justice, Defense, State, and the Office of the Science Advisor), Welsh sent Dutton in the White House some general conclusions. He believed "the majority, including myself," seemed to favor an organization with a joint government-private board of directors. He did point out, however, that NASA and Defense "did not show great enthusiasm" for having government representation on a board of directors. In response to a question "as to the real purpose" of their discussions, Welsh also clarified that they were examining "practical possibilities" so that they could be prepared if the proposals from the FCC's Ad Hoc Carrier Committee "did not seem appropriate."[137]

But it seems clear that Welsh was already prepared to reject any proposal coming from this committee, partly because he had decided on the need to involve Congress. Welsh informed Dutton, "It would probably be necessary to obtain legislation" if they decided to set up an organization for "owning the satellite system." Involving members of Congress would be necessary because they had "expressed such great interest in the matter" and would "want to have a hand in establishing a basis for its structure." And although "there was considerable discussion" about "what portion of the system, if any, should be owned by foreign governments or other foreign investors," Welsh reported that the committee had not come to any "firm conclusion." He emphasized that foreign ownership of overseas ground stations "was expected" and that "non-discriminatory access to and use of the system by foreigners was clearly a part of the President's policy paper."[138]

Decision to Seek Legislation

After the FCC's Ad Hoc Carrier Committee published its final conclusions on October 12, Welsh met with the Space Council agencies, and they all generally agreed that it was "inadequate." The committee recommended the establishment of a nonprofit corporation that would not own satellites or ground stations but would "provide minority representation to the U.S. Government on its directing Board." Only international carriers would participate in satellite ownership. But, according to the Space Council, the FCC acknowledged that the corporation and "the rest of the system would be substantially owned by AT&T." This was unacceptable to most agencies on the Space Council, especially Justice and State. The State Department was also critical of the report for being "completely silent on the question of how participation would be extended to additional countries." The agency continued to favor a "joint international undertaking." The Justice Department agreed but also argued that fundamentally, "this was a legislative problem; and action should not be taken by the FCC but must be taken by the President."[139]

The two most important officials involved with satellite communications policy in the Justice Department were Nicholas deB. Katzenbach, the assistant attorney general in the Office of Legal Counsel, who later became attorney general under President Johnson, and Lee Loevinger, chief of the Antitrust Division, who was later appointed to the FCC from 1963 to 1968. The son of a district court judge in Minnesota, Loevinger was serving as an assistant justice on the Minnesota Supreme Court when Kennedy appointed him in 1961. He was known for holding strong views on the need for antitrust enforcement. During his job interview, he told Robert Kennedy, "I believe in antitrust almost as a secular religion." And he had testified to a congressional committee a few years earlier that "the problem with which the antitrust laws are concerned—the problems of distribution of power within society—are second only to the question of survival in the face of threats of nuclear weapons in importance for our generation."[140] Loevinger was especially interested in issues involving science, technology, and law. He had majored in science as an undergraduate and helped establish and was an active member of the Science and Technology section of the American Bar Association.[141]

Katzenbach, a Princeton graduate and Rhodes scholar who taught at Yale and the University of Chicago, was the son of a New Jersey state attorney general. Katzenbach's *New York Times* obituary called him "one of the 'best and brightest,' David Halberstam's term for the likes of Robert S. McNamara,

McGeorge Bundy, Walt Rostow and other ambitious, cerebral and often ide-alistic postwar policy makers who came to Washington from business and academia carrying golden credentials." When Katzenbach's friend Byron R. White, the deputy attorney general under Robert Kennedy, was appointed to the Supreme Court in March 1962, Katzenbach replaced him as the number two official in the Justice Department. He played a central role in many of the major political decisions of the Kennedy and Johnson administrations, in-cluding passage of the 1964 Civil Rights Act and the Voting Rights Act of 1965. When Governor George Wallace refused to admit African Americans to the University of Alabama in 1963, Katzenbach confronted him on the steps of the main campus building and forced him to back down.[142]

Katzenbach encouraged the administration to seek legislation for satellite communications. According to an official account of a Space Council meet-ing in November, "He felt that the pressures were building up on Capitol Hill on both sides of the question, as well as within the Administration and that speed was of the essence." Although NASA had earlier favored the status quo, by November, key officials agreed that "legislative guidance" was needed.[143] Arnold Frutkin, director of international programs at the space agency, was especially "concerned about the possible disappointment of some small for-eign countries when they learn that access for them may be delayed for years" if the US government adopted the FCC carrier committee's plans.[144]

By mid-November, key White House aides had also concluded that new legislation was necessary. Frederick Dutton increasingly worried about the political implications of satellite communications. He warned the president on November 13 that "substantial Congressional and press concern continues over AT&T's potential strangle hold over communication satellites." Dutton informed the president that Senator Robert Kerr was preparing his "own leg-islative recommendations, so the entire matter will undoubtedly come to a head during the coming Congressional session."[145]

John Johnson, NASA's general counsel, helped write the satellite commu-nications bill Kerr introduced in the Senate. The head of NASA, James Webb, had asked Johnson to do this because Webb had previously worked for Kerr, and they were "good friends." Johnson recalled that Webb had called him to his office one day to meet with the senator. The two men had been discussing the "chaotic situation" involving satellite communications policy, and they felt "nothing was being done." Johnson had already been considering the need for legislation and "playing around" with ideas on his own, including a possible analogy with the legislation establishing the Union Pacific railroad. Kerr had an added incentive to introduce a bill because this would give the committee he chaired, the Space and Aeronautics Committee, "clear cut jurisdiction over the matter." John Johnson wrote the bill for Kerr, and then, according to Johnson, the senator delayed introducing it for "six weeks or so." During this time, beginning in the second half of November, Welsh convened several interdepartmental meetings with representatives from different government agencies, including John Johnson, which collectively drafted the administration's bill. But Johnson had not told other government officials about his role in helping to write Kerr's. When Kerr finally introduced his bill in the Senate, Johnson had to explain to other officials in the executive branch, especially Welsh and Katzenbach, why the draft of the administration bill contained similar language. Most important, he had to convince them, apparently to their satisfaction, that he had not leaked information to Kerr.[146]

The Kennedy administration introduced its own bill in Congress in early February 1962. Welsh had consulted with multiple agencies, including State, Justice, Defense, NASA, FCC, the USIA, the Bureau of the Budget, the Office of Emergency Planning, and the Science Advisor's Office. He did not bring the draft to a "formal vote of the members of the Council" but presented the draft document as a "well coordinated, but not formally agreed to, paper for such further action" as the president believed was "required and merited."[147] He also emphasized that it was consistent with the president's policy statement on satellite communications of July 24, 1961. Had there been a formal vote, NASA and the FCC likely would have voted against the administration's draft bill. Webb, as we have seen, had been working behind the scenes to support Kerr's legislative efforts. The FCC, along with the international communication carriers, especially AT&T, liked Kerr's bill more than the president's. AT&T had lobbied heavily for a bill favoring the international carriers. Welsh later recalled that "thirteen or fourteen Vice Presidents" from AT&T had come to his office to discuss the legislation: "They were trying to bring as much pressure on the drafting end and on the Hill and every end of the thing."[148]

The major difference between the two bills involved ownership of the new private commercial organization that, in both cases, the federal government would establish. Kerr's bill limited ownership to communications carriers and set the price of stock very high—one million dollars per share. Since AT&T was the wealthiest carrier and already controlled most international communications, it likely would become the largest shareholder in the corporation proposed by Kerr. Welsh believed his bill was "so blatantly a pro one company bill that it got very little support—in his Committee or from any of the members of the Congress."[149] The president's bill, by contrast, called for broad ownership, including not only common carriers and other companies but also all individuals interested in purchasing stock. To prevent monopoly control by one company, the president's proposed legislation also placed limits on the amount of stock an investor could own.

Katzenbach—one participant called him the "most effective person" assisting Welsh—had originally proposed the idea of allowing the public to subscribe to stock in the proposed corporation. Attorney General Robert Kennedy liked the idea, as did members of Congress, who, according to Welsh, "were very pleased that they were voting for the general public. So, it sold very well." Katzenbach and Welsh were concerned about protecting the public from the "risk nature of the investment," especially less sophisticated and poorer citizens. They decided to set the stock price relatively high, at a thousand dollars per share, to prevent poorer people excited about the new venture from potentially losing money if it did not work out as planned.[150] But when they sent the draft bill to President Kennedy, he decided that the price of stock should be set lower, "fairly low so that a lot of people could buy it."[151]

Compromise, Filibuster, and the 1962 Communications Satellite Act

After the Kennedy administration introduced its own legislation, Kerr decided to negotiate a compromise. Katzenbach received a telephone call from the president in February asking him to meet the senator for lunch at the Capitol. Katzenbach had taken the lead along with Welsh in presenting the administration's views to congressional committees, including Kerr's. The phone call occurred immediately after he testified in Kerr's committee. Katzenbach later recalled that they "went at each other fairly well in the testimony." The president told Katzenbach that Kerr was impressed with his performance and thought he was knowledgeable on the subject. During lunch, Kerr told Katzenbach that he did not want to oppose the president on the legislation, especially because he had recently worked against him on tax legislation. He explained his strong belief that the enterprise would fail if the carriers were not given a central role.[152]

Katzenbach and Kerr worked out a compromise that allowed the carriers to own 50 percent of the stock but also required public directors appointed by the president for the board of directors. And as part of the agreement, Kerr would be able to offer amendments that would affect the supervisory role of the government.[153] Katzenbach believed that Kerr forced AT&T to accept the compromise. Although he did not have direct knowledge of Kerr's interaction with the company, Katzenbach assumed that the senator, whom he characterized as "almost as powerful as the rest of the Senate put together at that time," had "jammed this down their throat." Katzenbach expected the presidential directors to look out for the public interest or uphold government policy rather than serve the interests of the dominant carriers, which would be both consumers of satellite communications and competitors controlling alternative technologies, especially undersea telephone cables.[154]

The idea to have some of the directors appointed by the president was first suggested by antitrust chief Lee Loevinger, who worried that AT&T would manage to gain control of the new corporation by buying up all the publicly owned stock. Attorney General Robert Kennedy often invited all the senior officers in the Department of Justice to have lunch with him one day a week in a private dining room on the fifth floor of the department, and Loevinger suggested this solution to Katzenbach during one of these get-togethers. Specifically, Loevinger thought public directors would "leaven the interest of those with proprietary interests," functioning as an internal control on the corporation by contributing to a system of internal checks and balances, or a form of competition.[155] These internal control mechanisms would supplement the external controls represented by federal agencies tasked with specific oversight responsibilities.[156] Loevinger believed the internal controls were necessary because traditional external controls would not be sufficient to prevent the disadvantages of monopoly.[157]

Other members of Congress submitted their own satellite communications bills, but the most important alternative to the administration's revised bill, now also supported by Senator Kerr, came from the liberal faction who believed the government should not turn over the results of its research to a private company that would effectively have monopoly control. They also generally believed that the only way to avoid AT&T dominance was to establish a government-owned system modeled on the Tennessee Valley Authority. And they worried that the president's bill would give the new corporation authority to negotiate international agreements that was properly the responsibility of the State Department.

The most important leader in the Senate of the liberal opposition to the president's bill was Estes Kefauver from Tennessee, chairman of the Antitrust and Monopoly Subcommittee. Although he was especially famous for chairing during the early 1950s a special Senate committee investigating organized crime, which became known as the Kefauver Committee (Special Committee on Organized Crime in Interstate Commerce), he also was one of the most important critics of economic monopoly during the postwar period. His antimonopoly subcommittee had one of the largest staffs of any committee on Capitol Hill and, according to the staff director, included "probably the finest antitrust economist in the country" during the early 1960s, John M. Blair, former assistant chief economist of the Federal Trade Commission.[158] Kefauver's subcommittee investigated a large number of industries during the late 1950s and early 1960s, including steel, auto, and electrical manufacturers, as well as the milk and drug industries and professional sports. Transcripts of hearings came to some 18,000 pages in twenty-nine volumes.[159] Although the hearings did not generally lead to major changes, powerful corporate leaders were forced to answer probing questions about their businesses.

Historians have pointed out that Kefauver was generally not in tune with postwar liberalism. According to Daniel Scroop, he was rooted more firmly in the New Deal and in a "populist-progressive tradition" that had its origins, "at least partially, in agrarian distrust of concentrated economic and political power." Most liberals grudgingly accepted the increasing tendency for big business to become entrenched through mergers and acquisitions, which had accelerated during the Eisenhower administration. Rather than focus on eliminating monopolies, liberal economists often tended to emphasize the role of organized labor as a "countervailing force against big business."[160] But the Kennedy administration could not easily ignore Kefauver's energy and crusading spirit when he opposed the president's satellite communications legislation, especially since it was focused not on a relatively mundane technology but on a glamorous space-age innovation.

Three of the strongest supporters in the Senate for Kefauver's bill proposing government ownership of the planned satellite communications organization (S. 2890), which he introduced in Congress in late February, were Albert Gore from Tennessee, Wayne Morse from Oregon, and Russell Long from Louisiana. They all served on the Foreign Relations Committee, and Gore was particularly concerned about the State Department's role being weakened in the president's amended bill.[161] Representatives William Fitts Ryan of New York and Frank Kowlaski of Connecticut introduced similar bills supporting government ownership in the House of Representatives (H.R. 9907 and H.R. 10629).[162]

To understand the motivations of this liberal faction in Congress, it is important to understand the significance for them of the atomic energy debates during the 1950s, and their understanding of the meaning of the new technology. Many government officials, including the members of the liberal faction, viewed satellite communications as a radical innovation similar to atomic energy and rejected the FCC understanding that communications satellites were merely an extension of traditional communications. Significantly, for purposes of organization, the government tended to put space and atomic energy together; for example, during 1961, the State Department had an Office of Special Assistant for Atomic Energy and Outer Space, and the Space Council included the head of the Atomic Energy Commission as one of its official members. To the liberal faction, the satellite communications legislation looked similar to the Eisenhower administration's 1954 Atomic Energy Act, which they had vigorously fought. The atomic energy legislation had opened up the government monopoly of atomic energy to private development. Senators Gore, Kefauver, and Morse had unsuccessfully tried to block the legislation, arguing that the Eisenhower administration was guilty of giving away to private industry billions of dollars of government-funded research. The two Tennessee Democrats had also been upset because of Eisenhower's efforts to use the Atomic Energy Commission to subsidize a private power company and undermine the Tennessee Valley Authority. The filibuster by Gore and the liberal faction against the satellite communications legislation was inspired by the memory of losing this earlier fight over another cutting-edge technology funded by the federal government.[163]

Senator Kerr's Space Committee was the first committee to report on the president's legislation. His committee added several amendments with its favorable report on April 2, including some that were clearly meant to weaken the regulatory role of the president and key government institutions. Especially important were the amendments that attempted to make sharp distinctions between foreign affairs and business considerations. For example, one amendment removed the original language giving the State Department the power to "conduct or supervise" negotiations with foreign entities and replaced it with language giving the department a purely "advisory" function. Another amendment seemed to water down the ability of the FCC to ensure that decisions about technical specifications involving system choice also considered the broader geopolitical concerns of the government.[164]

In the House, Representative Oren Harris introduced, also on April 2, a bill (H.R. 11040) identical to the amended legislation that came out of Kerr's committee (S. 2814). The House Committee on Interstate and Foreign Commerce gave a favorable report with only a few minor, nonsubstantive changes. The Rules Committee then decided to introduce the legislation on the floor of the House. Congressmen William Fitts Ryan attempted unsuccessfully to add an amendment allowing government ownership. Generally, the bill was favorably received by the full House, which voted overwhelmingly in support of the legislation on May 3. Only 9 members voted against H.R. 11040; 354 voted in favor.[165]

The president's amended bill was sent back to the Senate to be taken up by the Commerce Committee, which had already, in late April, considered the amended bill that had come out of the Senate Space Committee. The head of the committee, Senator John Pastore, generally favored the president's original bill, and his committee added amendments intended to give back some of the original authority for the president and federal agencies that had been weakened by Kerr's committee. This bill was introduced on the floor of the Senate on June 14, and then bitterly attacked for an extended period by the liberal faction led by Senator Kefauver. To give time for other important legislation, the Democratic leadership removed the bill from consideration for five days beginning on June 21. When the leadership attempted to reintroduce the bill on the floor of the Senate, the opponents launched a filibuster that held up business in the Senate for six days. At that point, Democrats and Republicans agreed to introduce a cloture motion to end debate, but before taking this drastic action, they agreed to a compromise with the opponents that involved having the Foreign Relations Committee hold hearings to examine the foreign policy implications of the proposed legislation. The committee was expected to refer the bill back to the full Senate by August 10.[166]

The filibuster against the president's satellite communications legislation was unique because it almost entirely involved liberal Senators mainly from the North (including northern border states such as Tennessee) who were motivated not only by the legacy of the atomic energy debates but also by concerns about potentially allowing AT&T to extend monopoly control into outer space. During this period, filibusters were generally uncommon; however, the exception was the use of the filibuster by conservative southern Democrats to try to stop civil rights legislation. Bernard Fensterwald, the staff director of Kefauver's antitrust subcommittee, pointed out that filibusters "were for civil rights and were always run by Southerners." According to Fensterwald, "We organized the first liberal filibuster. There had never been one" before this. The approximately twenty-one liberal Democrats who participated in this filibuster "were the same people that had ranted and raved against filibuster for years."[167] They were fortunate to have the support of Senator Long, a southerner with extensive experience running filibusters. "He was," according to Fensterwald, "the only one who knew anything about running a filibuster." He had been in "20 or 30 different filibusters." To run a filibuster, senators needed to follow specific procedural rules: "It's not all that simple, because you can break a filibuster unless it's properly done, and he taught us how to do it."[168] Long "got these 22 Senators in a room and witchheaded them for about two or three hours about how you run a filibuster, and how if they were going to do it they had to do it; and you know, no fiddle faddle."[169]

Since three of the main opponents of the president's bill—Wayne Morse, Albert Gore, and Russell Long—were on the Foreign Relations Committee, they hoped that the compromise allowing their committee to conduct further hearings would give them an opportunity to substantially change or put a stop to the legislation. But the other members of the committee were not as concerned about the bill and were not convinced by the extensive testimony and questioning organized and led by the three senators. The administration sent top officials from executive branch agencies to testify, including the secretary of state and the secretary of defense. And the FCC put aside its own opposition and for the first time strongly supported the administration's bill. Dismissing the objections of the three opponents, the committee reported the bill favorably to the full Senate on August 10.[170]

The main opponents in the Senate relaunched their filibuster as soon as an attempt was made to reintroduce the bill on the floor of the Senate, but the leadership introduced a cloture motion, and the Senate voted for cloture (for the first time since 1927), with sixty-three senators in favor and twenty-seven against. Although the opponents managed to have a few of their own amendments included in the final Senate legislation, these were mainly minor adjustments that did not change the central issues. The Senate then passed the amended legislation on a sixty-six to eleven vote. The House of Representatives voted in favor of the final bill on August 27, and the president signed the legislation on August 31.[171]

The large majority approving the legislation in the final votes in the House and the Senate underscored that the liberal faction's desire to have legislation passed establishing government ownership for the planned satellite communications organization was clearly unrealistic. But Fensterwald argued that the senators involved with the filibuster had come much closer than the final vote indicated to blocking the legislation. He blamed the defeat on Senator Gore: "But except for slightly bad nerves on the part of Albert Gore it would have been defeated."[172] According to Fensterwald, the best time during this period to start filibusters had been May or June because legislation would have to be called off if organizers could sustain the effort through June 30. That date was crucial because "all sorts of catastrophic things happen if certain Acts are not renewed." Congress normally automatically renewed important government programs by this date "every year, year after year," including programs having "to do with veterans['] preference and funding of the International Bank and the Treasury's issuing money." If the funding was not extended by this date, he recalled in an interview, "all sorts of things grind to a halt."[173]

The organizers of the filibuster believed at the time, according to Fensterwald, "All we've got to do is get up to June the 30th and everybody will call this legislation off; we will have won." He remembered that it had been "working splendidly and we were coming right up to the 30th of June and everybody was fairly rested and so forth, and Albert Gore got terribly patriotic, and he says, 'We've got to give up the Floor long enough to get these technical extensions.' " Gore "got enough support that that's what happened," according to Fensterwald. To continue the filibuster they would have had to "start all over again," and because the pressure of the deadline no longer existed, they would have had to "argue another six months before" reaching a point where they could win. Gore and the other senators were under tremendous pressure from the president and the majority leader, "and everybody was yelling at them."[174] Fensterwald's theory assumes that the cloture votes would have been defeated if Gore had not caved in to the pressure earlier in June; this is impossible to prove, but the filibuster clearly came closer to succeeding than has usually been acknowledged.

And although Gore and the other liberal members of the Foreign Relations Committee did not succeed in gaining support for far stronger language in the final bill mandating a central role for the State Department, they did succeed in getting their point of view in the public record. The State Department's role had originally been weakened partially because of the performance of the undersecretary of state for political affairs, George McGhee, during testimony in Senator Kerr's committee. Another State Department employee, William Gilbert Carter, who was hired in 1962 immediately before the filibuster to help deal with the international negotiations for the planned global system, recalled that Kerr "had really chopped him [McGhee] up in small pieces during the testimony and had left the State Department in a weak position under the report language that came out of the Senate and the draft language of the bill;

weak vis-à-vis, the to-be-formed Communication Satellite Corporation." Because of McGhee's weak performance, the "activist role" for international negotiations "was clearly focused on the Communications Satellite Corporation rather than on the [State] Department." The additional hearings in the Foreign Relations Committee, which included the testimony of Secretary of State Dean Rusk, according to Carter, "provided a very, very useful legislative history from the standpoint of the State Department aimed at, if you will, readjusting the balance between Comsat and the State Department which had been quite badly destroyed by McGhee's testimony." The general language in the final legislation signed by the president was open to interpretation; this meant that much depended on the specific decisions of government officials tasked with working with Comsat in establishing the international consortium, and their decisions in turn depended on the legislative history coming out of the filibuster.[175]

The final legislation set aside 50 percent of the stock in Comsat for the international carriers, but also mandated broad ownership, including distribution of the remaining stock "in a manner to encourage the widest distribution to the American public." No communications carrier could have more than three members (out of a total of fifteen) on the board of directors. Altogether, six members would represent the communications companies, six would represent the public shareholders, and the president would appoint three. Different agencies of the government also had a role in regulating Comsat, including the State Department.

One interpretation of the significance of the 1962 legislation is that it represented a step on the way to the eventual breakup of AT&T in 1982, which resulted from an antitrust suit initiated in 1974 by the US government against the company. Before this, AT&T had operated according to the terms of a consent decree issued by the Justice Department in 1956 that allowed the Bell System to have a monopoly as long as it restricted its activities to the national telephone system and serving the government. But during the late 1950s and early 1960s, the FCC made some key decisions that began to restrict AT&T's activities and introduce competition.[176]

Members of Congress who were concerned about AT&T gaining control of the planned satellite communications system also discussed some of the broader issues that at least indirectly influenced the FCC's efforts to restrict AT&T during this period. Representative Emanuel Celler of New York called AT&T an "old offender," which "does not come before the bar of the Congress with clean hands." Celler complained that the commission had allowed AT&T

to enjoy a rate of return far in excess of that to which it was entitled. He believed the company had "overcharged the American public by nearly a billion dollars" over a seven-year period beginning in 1955. "Divine guidance," according to Celler, "would be necessary to regulate AT&T if it is permitted to expand its domain into space."[177] The FCC confirmed that at least in some cases, it had not rigorously regulated AT&T because it lacked the funds and personnel to do so.[178] Testifying before Congress in August 1961, Assistant Attorney General Loevinger revealed that the Justice Department had been actively investigating whether AT&T had used its monopoly position in the past to stifle innovations, such as automatic switching equipment and mobile handsets.[179] Also during this period, the Commercial Telegraphers' Union called on President Kennedy to take a stand against the "unfair, ruthless competition of this all-powerful communications giant [AT&T]," which had "destroyed" the jobs of many union members as it steadily moved into the telegraph field.[180] Finally, representatives of rural telephone cooperatives urged members of Congress to question whether the same company that had resisted serving rural areas of the country would likely establish a communications satellite system serving all areas of the globe.[181]

Those members of Congress who favored government ownership of the planned satellite system believed that, especially in the case of satellite communications, domestic policy had to consider global concerns. And this helps explain their fears about AT&T control. They argued that the Soviets would use a decision to give private companies free rein as material for anti-American propaganda. Senator Wayne Morse reminded his colleagues that US telephone companies operating in Latin America that had recently refused to serve unprofitable areas had played "right into the hands of the Russians."[182]

Many government officials who promoted satellite communications predicted that one of its most valuable uses would be in support of television broadcasting, especially to non-Western countries. Members of Congress who worried about AT&T dominating satellite communications argued that the nation should not allow private interests to control "this vast new medium of communications" since it would be a "great boon . . . to the capacity of the United States to project its people, its way of life, its programs and policies to the one-third uncommitted part of the world and acquire their loyalty and friendship."[183] Edward R. Murrow, the head of the USIA, pointed out that he would never be able to use satellites for broadcasting if AT&T controlled the system and charged the same overseas rates it charged for the use of its cables. The agency would need to spend approximately $900 million each year, he estimated, to conduct effective global television broadcasting.[184] Senator Morse questioned why Congress should support commercial domination of a system "in which the American taxpayer is going to have to pay a huge sum of money to export our ideology of freedom abroad to these underdeveloped areas and thereby guarantee a nice big, fat, juicy profit to AT&T and other corporate entities."[185] When members of Congress thought about using satellites for broadcasting television, they were particularly fascinated with "direct broadcasting," which led to predictions of "a day when in the poorest places in the world there will be screens which themselves can take right off of the satellite President Kennedy's broadcasts [and] press conferences." Commentators acknowledged that communications experts had emphasized that this form of broadcasting would take powerful satellite transmitters and other equipment not yet available, but they thought it would still likely occur.[186]

Morse and others warned that the system AT&T was promoting would not satisfy the national objectives for true global coverage. Senate Russell Long of Louisiana thought NASA's arrangement to launch AT&T's *Telstar 1* satellite might prove to be a "foot-in-the-door contract," locking the United States into a commitment to an inferior system. Members of Congress who favored government ownership found the testimony of officials from Hughes and other companies about the inherent superiority of the geosynchronous design convincing; they were particularly worried about the effect on the country's image if the Soviets put up a synchronous system first.[187] Fred Adler, manager of the Space Systems Division at Hughes, emphasized that non-Western countries would not be able to afford the complex and expensive ground stations used with AT&T's Telstar satellite. "Newly emerging nations, like Ghana," with relatively small traffic requirements, he pointed out, "would require investments in excess of $10, possibly $30 million" to put in a "full Telstar type of ground station." Adler predicted that AT&T's medium-altitude system would most likely result in "one large gateway station in Africa, like in Johannesburg, and you would have to route all the traffic to this one station and take it from there." He asked, rhetorically: "Suppose we go ahead with a medium altitude system which directly serves only the 'have' nations. Then let the Soviets follow with a stationary multiple-access system which ties all the 'have not' nations to the USSR. Who will have won this vital race for men's minds?"[188]

Although the restrictions placed on AT&T's work in satellite communications by the Kennedy administration could be viewed as a further example of the general trend driven by domestic antitrust concerns, the analysis in this

President John F. Kennedy signs H.R. 11040, the Communications Satellite Act of 1962. Immediately to the right behind him is Vice President Lyndon B. Johnson. Photograph by Abbie Rowe, 1962.

chapter of the policy debates leading to the passage of the Communications Satellite Act of 1962 provides a deeper understanding of the reasons government officials and legislators worried about monopoly control by AT&T. In the context of total and global Cold War, traditional domestic antitrust concerns became linked to geopolitical imperatives. But although members of Congress and officials in the Kennedy administration were motivated, in general terms, by the Cold War in pushing for a global satellite communications system that would not be dominated by AT&T and that would be open to the involvement of all countries, they did not spend much time discussing specific details about foreign involvement. The State Department and key members of Congress were especially concerned about ensuring that the US government, rather than the new American corporation established by the 1962 Communications Satellite Act, played a dominant role in negotiations with other countries, but they did not discuss the idea of a global system with other countries until after the legislation was passed. Particularly because the 1962 act was open to interpretation on many issues, the planned international consortium was largely a blank slate until key State Department and Comsat officials made initial efforts to consult with other countries beginning in late 1962 and early 1963.

Global Satellite Communications and the 1963 International Telecommunication Union Space Radio Conference

Before the global satellite communications system envisioned by the Kennedy administration could be established, the United States had to convince the 1963 Extraordinary Administrative Radio Conference (EARC) of the International Telecommunication Union (ITU) to set aside large blocks of the radio spectrum for satellite communications. Informally known as the Space Radio Conference, the event was the first meeting called by the ITU exclusively to decide about allocating frequencies for all types of space activities. In this chapter, I first examine the crucial national security context that helps explain the US position at the 1963 meeting and then explore the specific decisions and actions of US officials before and during the meeting.

To understand the general success of the United States in convincing countries to agree to its proposals for space frequencies at the crucial 1963 ITU meeting, it is important to understand not only the significance of changing US-Soviet relations but also the specific strategies used by American diplomats and other government officials. Particularly significant was the use of spectacular demonstrations, diplomatic "missionary" activities in preparation for meetings, and a form of boundary work to finesse the interrelated geopolitical and technical issues central to the work of international governmental institutions such as the ITU. An analysis of these strategies allows us to better understand the negotiations leading to decisions about radio frequency standards and how these decisions provided an essential foundation for the satellite communications global infrastructure project. In addition, these strategies also deepen our understanding of how the United States worked to shape major international institutions during the Cold War as part of a broader soft-power objective to gain influence internationally, but more specifically to keep countries in the Global South from going communist.

National Preparedness and the Cold War

US efforts leading to the 1963 ITU Space Radio Conference were motivated by both national security and commercial considerations. Although the federal government established Comsat and put in place measures to guarantee strong oversight, it was a commercial company that sold shares to the public. Thus, setting aside frequencies through international agreements was necessary to create confidence and attract investors potentially worried about the future development of the global system. General national security concerns involving a symbolic and material struggle over hearts and minds, as well as military preparedness, also drove US efforts leading to the Space Radio Conference. Specifically, the implications for electronic communications of the crises involving Cuba in 1961 and 1962 were an important factor influencing the United States. Difficulties with communications among agencies of the US government contributed to the failure of the Bay of Pigs invasion in 1961. And during the Cuban missile crisis in October 1962, President Kennedy was unable to communicate "in a timely fashion" with multiple foreign leaders, including the heads of Latin American countries and of the Soviet Union.[1]

The Cuban problems focused the attention of the US government on the need for "a reliable, secure, flexible link for Presidential and command control communications to remote areas of the world and to meet crisis emergencies."[2] A National Communications System Working Group established in late August 1962 by National Security Advisor McGeorge Bundy and Secretary of Defense Robert McNamara led to the organization of a Subcommittee on Communications by the National Security Council.[3] Named the Orrick Committee after the chair, William H. Orrick Jr., the deputy undersecretary of state for administration, it recommended in May 1963 that the president integrate separate government communications systems into a single entity. The president responded by establishing the National Communications System (NCS) in August.[4] The NCS not only facilitated worldwide US military communications but also provided "24-hour, seven-days-a-week communications response capability" at key diplomatic posts, particularly in Latin America.[5]

Government officials involved in organizing the NCS viewed communications satellites as a new technology that would supplement or provide redundancy to existing networks, mainly international radio and undersea cables. Satellite communications was especially important because of its potential ability to operate, unlike international radio networks, even in the event of a nuclear attack on ground installations. The organizers of the NCS hoped that they would be able to lease channels from the global system Comsat was established to organize (the system that became known as Intelsat).[6]

As major users of government communications networks, the Department of Defense (DOD) and the military agencies played a central role in the organization of all aspects of the NCS, including the possible use of communications satellites.[7] Well before the establishment of the system, the military had become deeply interested in communications satellites. The DOD was committed not only to developing separate military systems for classified communications but also to influencing the character of the planned commercial system. The military's interest in satellite communications thus became an important factor driving US planning for the 1963 Space Radio Conference. The importance of the military for space radio frequency planning in general should not be surprising given that "substantially more than half of the 90 payloads successfully put into Earth orbit in 1962 were military satellites" and that "at least two-thirds" of the launchings during that year were classified as "secret."[8]

The military was especially interested in communications satellites for their ability to facilitate access to remote areas of the world through mobile Earth stations. Because of the central role of the DOD in the organization of the NCS, the use of "mobile stations which could be quickly transported to any part of the world, set up and be in operation with high quality communications in a very short period of time" became one of the system's key requirements.[9] The DOD emphasized that it would "apply the same policy to satellite communications as to other long-line communications," that is, it expected military departments to "use US commercial facilities and provide exclusive military facilities only when required to meet those military needs that cannot be satisfied by commercial facilities."[10]

But the DOD was not simply interested in passively using commercial communication systems; it also sought to actively influence their technical development to make sure they would be militarily useful. In September 1961, the department announced that it wanted the proposed commercial global communications satellite system to "be capable of providing sufficient flexibility to provide for the wide support of military force deployment including certain forces under mobile conditions," especially in "those portions of the world having non-existent or primitive communications."[11] This was justified, according to Harold Brown, the director of defense research and engineering in the Office of the Secretary of Defense, because of the critical demands of the Cold War. "Our domestic and international telecommunication systems are critical factors both in our military posture and in the Cold War struggle," he argued, "and, indeed, throughout the whole spectrum of conflict. We cannot today consider our communication systems solely as civil activities merely to be regulated as such, but we must consider them as essential instruments of national policy in our struggle for survival."[12]

The DOD was also interested in building separate military systems for classified communications that could not go over commercial facilities. Military officials liked the prospect of geosynchronous satellites rather than lower-orbiting medium-altitude satellites because their simpler, and potentially portable, ground stations seemed to fit better with the military goal of providing communications to remote regions.[13] Beginning in 1960, military agencies (mainly the army and the air force) spent more than $160 million on the sophisticated geosynchronous system called Advent before the DOD canceled the project two years later when it became clear that the complex technology was not yet available and that Congress was not willing to accept further "severe financial overruns."[14]

Late in 1962, the department reoriented the program by pursuing both geosynchronous and complex nonsynchronous (medium-altitude) designs in the long term while focusing in the short term on the establishment of a simpler and more realistic medium-altitude communication satellite system.[15] These new plans remained open, however, during the period leading up to the November 1963 ITU Space Radio Conference. Another complication was that as early as May 1963, the DOD and other government agencies considered the possibility of working with Comsat in designing the planned global communications satellite system (either geosynchronous or medium altitude), so it could be used by both US military agencies and the planned international satellite consortium.[16]

Establishment of Comsat

Fifteen officials appointed by President Kennedy on October 15, 1962, formally incorporated Comsat in the District of Columbia on February 1, 1963.[17] The Department of Justice reserved the name Communications Satellite Corporation (Comsat) at the Office of the D.C. Superintendent of Corporations.

The president asked Philip Graham, the owner of the *Washington Post*, who was a close friend and supporter of Kennedy, to serve as chairman "until such time as the group can adopt its own rules of procedure and make its own choice of a chairman."[18] While operating the newspaper during the 1960 presidential election, Graham had advised Kennedy, helped write some of his campaign speeches, and played a key role in the choice of Lyndon Johnson to run with him as vice president. A former Comsat employee described Graham as a "very gung-ho," "freewheeling" type who tended to make decisions on his own and then seek approval later from the other incorporators.[19] Despite this criticism, one of the other incorporators pointed out the difficulty in getting fifteen people to make decisions, concluding that had it not been for "Phil's initiative we'd never [have] gotten off the ground."[20]

The incorporators largely represented industry and banking; five were practicing lawyers, and—with the assistance of another Kennedy supporter, Lloyd Cutler, and his law firm of Wilmer, Cutler, and Pickering—they helped develop articles of incorporation and bylaws for the new corporation.[21] The incorporators included Sam Harris, the vice chair, a New York corporate lawyer with experience in investment banking and securities regulation, who had also been "active in the Kennedy campaign"; Sidney J. Weinberg, partner in Goldman Sachs, New York City; Edgar F. Kaiser, president of Kaiser Industries Corporation, Oakland, California; David M. Kennedy, chairman of the Continental Illinois National Bank and Trust Company of Chicago; Leonard Marks, a Washington, D.C., lawyer with expertise in communications law and the regulatory work of the Federal Communications Commission (FCC); and Leonard Woodcock, UAW-CIO vice president, Detroit, Michigan.[22]

Along with setting up the company, the incorporators needed to come up with initial equity capital and prepare for the public stock offering. As one of the early Comsat employees argued later, because the 1962 Communications Satellite Act did not authorize any funds for the new corporation, it was basically "insolvent at the outset."[23] Initial capital came from the incorporators, who each invested one hundred dollars.[24] Kennedy then used his connections as a major banker to arrange for loans from ten separate banks, each for five hundred thousand dollars.[25] The company did not have any assets or security at that point, but they did have, as one of the incorporators later pointed out, "a license from the United States government to organize."[26]

Before stock in the corporation was publicly offered to investors in early 1964, the incorporators interviewed candidates for positions in the company and made initial major hires, most important appointing Joseph Charyk as the chief operating officer (COO), or president, and Leo Welch as the chief executive officer (CEO). Charyk had a PhD in aeronautics from the California Institute of Technology, had served as the undersecretary of the air force, and had overseen "most of the Department of Defense's satellite programs," especially the highly classified reconnaissance satellites.[27] Phil Graham had been mainly responsible for recruiting Joseph Charyk, who had already signaled to the DOD that he planned to look for other opportunities outside government. Charyk's boss, Deputy Secretary of Defense Roz Gilpatric, initially informed Charyk about the potential position at Comsat and offered to recommend him to Graham. Charyk met with Graham several times in late 1962 and then with the board of incorporators in early 1963. When they offered him the position, they emphasized that they expected Charyk to be mainly responsible for developing "the concept for what kind of a system was needed and all of the matters relative to the establishment of a system." According to Charyk, the incorporators did not discuss such basic issues when they hired him. They were especially interested in finding a COO and a CEO who would "be complementary to each other," with the CEO having "extensive business experience, since one of the big problems would be raising the capital for the new company," and dealing with all the "financial contacts."[28] Leo Welch was in many ways an ideal fit, with international corporate and banking expertise, having served as chairman of the Standard Oil Company of New Jersey and before that as an international banker in Latin America.[29]

Charyk and Welch played key roles in hiring other early employees for the different divisions in the company. These included lawyers, notably the general counsel, Allen Throop, who had extensive business experience, particularly involving work with the Securities and Exchange Commission, and David Melamed, who came to Washington to serve as assistant counsel to Graham soon after the incorporators were appointed.[30] Employees working in technical divisions mainly came from private industry and the government, especially the military and NASA. Siegfried Reiger, a German rocket scientist who had worked with Wernher von Braun at the Peenemünde rocket and missile facility in Germany during the war, became the head of the Systems Division.[31] Before his Comsat appointment, Reiger had served as a senior staff member and head of communications satellite studies at the RAND Corporation. Soon after being hired, Charyk and Welch visited RAND to hear Reiger give a presentation on different satellite communications systems.[32] Sidney Metzger was appointed head of the Engineering Division in the middle of 1963. He came from RCA, where he had worked with the company's Relay communication satellite program. The satellite communications business was not large, he pointed out later; "it was just a handful of people, and I knew many of them." Not surprisingly, he hired other former RCA employees to work in his division.[33]

Several other employees came from the DOD. At least two people who had worked with Charyk at Defense came with him to Comsat: Lewis Meyer, who became the financial coordinator in June 1963, and Donald Greer, Charyk's executive officer, who assumed a similar role at Comsat.[34] Another former DOD employee, John Johnson, became the director of international arrangements and played a crucial role in developing agreements with other countries that led to the establishment of Intelsat in 1964. (He also served as the first head of the organization.) He and Charyk knew each other from the air force, when Johnson had served as the general counsel. One former Comsat employee, William Berman, thought Johnson and Charyk were excellent examples of the well-known group of whiz kids Secretary of Defense McNamara brought with him to the Pentagon: McNamara's "brilliant young guys in the Defense Department." After the air force, Johnson worked for NASA, where he served as general counsel and headed the international program, developing cooperative relationships with other countries. Since he had conducted similar international work at the air force, he was an excellent choice to oversee international activities at Comsat. Berman recalled that Johnson could be "unbelievably argumentative . . . and loved to hear the sound of his own voice," but he also described him as "absolutely brilliant, articulate, dynamic, [and] hard driving (in a nice way)."[35]

People accepted positions at Comsat for different reasons, but notably, in at least two cases, employees emphasized the appeal of being able to work for an enterprise that combined a goal of profitability with "a certain amount of altruism." Don Greer especially appreciated that work with Comsat led to a "certain feeling that you were doing things that were really important to the country and the world, that were needed, [and] that were going to do good things for people." It was, he argued, not like "just selling a different colored pencil or a better mouse trap that really wasn't needed and if you never had invented it or sold it, the world wouldn't have cared very much."[36] A former RCA employee who headed the Terrestrial Interface Department at Comsat, James Potts, expressed a similar idealistic appreciation about how his work would bring "communications to the world." He also appreciated not being involved in defense work, which had been his main responsibility at RCA. According to Potts, "I really didn't like the defense world . . . You had nice problems and very unusual engineering problems, and you could get excited about the problem, but you [couldn't] get terribly excited about the results in a lot of cases."[37]

Perhaps reflecting the unique origins of Comsat, the first building the company leased for its headquarters, beginning in early 1963, was equally unusual: a 1912 mansion called Tregaron located on "the old Evans Estate" in Northwest Washington, D.C. The mansion had been owned by Joseph E. Davies, US ambassador to the Soviet Union during the 1930s, who authored a book about his experiences, *Mission to Moscow*, and his wife Marjorie Merriweather Post, heiress to the Post cereal company.[38] Charyk's office was the master bedroom on the second floor up a "broad, winding staircase." His secretary's office was the adjoining dressing room. Welch's office was also on the second floor, "just off the foyer in a paneled library."[39] A former employee remembered that they placed the company's mimeograph machine on a "big black stove in the kitchen."[40]

Before the company could sell stock to investors, it needed to decide about the size of the offering. Graham's behavior during heated discussions about the size of the stock offering at early meetings of the board of incorporators increasingly seemed to show that he "was not well."[41] A combination of heavy use of alcohol and increasing evidence of bipolar disorder led to his resignation from the board in early 1963 and the appointment of Sam Harris as the new chair. Graham also began an affair with an Australian *Newsweek* employee, Robin Webb, during this period and threatened to leave his wife, Martha Graham, whose father had previously owned the *Washington Post*. Martha Graham sought professional help to treat her husband's mental illness during the first half of 1963, which included efforts to have him committed to psychiatric institutions, but tragically, in August, Graham committed suicide. Harris in many ways was the exact opposite of Graham. Charyk described Harris as a "very conservative, well-heeled business-type," who was "very soft-spoken, very deliberative, [and] very organized."[42]

The board of incorporators debated two basic views related to the size of the stock offering. Some members favored a "two-step" approach, which called for an initial relatively small public offering because of resistance from the carriers.[43] The 1962 Communications Satellite Act required issuing half the stock to international carriers, but at first all the carriers expressed strong reservations about making a major investment. Because of the uncertainties remaining in satellite communications, especially for achieving a commercially viable global system, the carriers preferred a smaller initial stock offering. Some of the incorporators sided with the carriers, notably John Connor, the president of Merck, who, according to Throop, "was very vehement that [Comsat] should make a twenty-five million offering" and then finance the balance "either with additional stock or borrowing or both."[44]

Other incorporators favored a "go for broke" approach, in which the company would make a large public offering, ideally big enough to support a demonstration of a commercially viable system. Especially because of Charyk's support, the incorporators voted in favor of this approach. The company conducted initial studies about the likely costs of different systems, notably medium altitude versus geosynchronous, and although this gave a general idea of potential costs, Charyk later argued that it was not "at all clear what amount of money would ultimately be required."[45] They did know that a workable geosynchronous system should be cheaper, simply because it would not need as many satellites. A medium-altitude system, should they decide to build it, would require a comparatively large amount of capital. The $200 million stock offering they chose was thus large enough to cover all potential system options, including a medium-altitude system. The "desire to lean a little on the heavier side" so they could "really demonstrate what a system could do" outweighed "original feelings" to settle for a lower amount.[46] The carriers had signaled that they would be willing to lend any additional funds needed after an initial smaller stock offering, but Charyk and others worried that this "would really give the carriers fantastic leverage on the company, and that would not be a sound foundation for the healthy growth of the company."[47]

AT&T's willingness to cooperate after initial skepticism also made it easier for Comsat to choose a larger stock offering. Since Comsat had been established by the government, and AT&T was a regulated monopoly that had to worry about hostile critics in Congress, the company needed to "look cooperative."[48] John Johnson argued that an additional reason for AT&T's cooperation was that the company was confident Comsat would fail if they chose to pursue a geosynchronous system, at which point AT&T executives believed Comsat would have to fall back on the medium-altitude system they had been promoting.[49] Initially, the FCC, which had to sign off on the decision about the size of the stock offering, was also concerned that $200 million might be too large but later backed down. According to a former employee at the commission, they believed it was important to allow Comsat "to have enough money for a worst-case basis rather than for a best-case basis."[50]

The prospectus for selling 10 million shares of stock in the company detailed thirteen risks investors needed to consider, including the experimental and untested nature of the enterprise; the inherent risk of launch vehicle failure; dividends on the common stock not being paid "for an indeterminate period"; the risk of "early failure of satellites"; uncertainties connected with the need to compete with undersea telephone cables; the need for international partners who had not yet been fully consulted; and the possibility that "other countries may have or may acquire the ability to establish and operate a satellite system which would complete with the system the Corporation proposes to establish."[51] Throop believed that these different risks, which the prospectus "spelled out" in detail, should have "scared the hell out of anybody who had any sense." But that did not happen.[52] The public "gobbled up the issue."[53] Consistent with the requirements of the legislation to distribute the public shares as widely as possible, individuals were limited at first to fifty shares, but by the time brokers actually offered up shares to investors, they reduced this number to an average of ten shares per customer.[54] The enthusiastic response was partly because the public viewed it as a glamorous new space-age technology. But Throop also pointed out that "sometimes the more speculative you make something, the more attractive it is."[55]

Based on the requirements of the 1962 Communications Satellite Act, the carriers would purchase 50 percent of the stock, or make a financial commitment of $100 million. AT&T came through with a large investment as promised, but the major surprise was the second largest international carrier, ITT. Charyk commented that although a "small actor in the thing," especially compared to AT&T, ITT still "came up with a major investment and that was a surprise to everyone."[56] The carriers together oversubscribed by 27 percent of their authorized $100 million. AT&T offered to purchase $85 million of stock, but because of the oversubscription, it needed to reduce its share to $57.9 million. ITT's share, the second largest, was $21 million.[57]

A new board of directors replaced the original board of incorporators at the first annual meeting of the shareholders on September 17, 1964. Based on the requirements of the 1962 Communications Satellite Act, the board consisted of fifteen individuals, with three appointed by President Lyndon Johnson (Frederic G. Donner, George R. Meany, and Clark Kerr), six elected by the communications common carriers that owned common stock (Eugene R. Black, Harold M. Botkin, James E. Dingman, Douglas S. Guild, Horace P. Moulton, and Ted B. Westfall), and six elected by the public shareholders (Joseph V. Charyk, David M. Kennedy, George Killion, Leonard H. Marks, Bruce G. Sundlun, Leo D. Welch).[58] In June 1964, the company awarded contracts for engineering design studies for three different communications sat-

ellite systems, including medium altitude and geosynchronous. Later that year
the company planned to enter into agreements to develop actual satellite systems, "of one or more types," with the expectation that it would make a final
decision on which one would be used with the operational commercial system
during the second half of 1965.[59]

Planning in the United States for the 1963 Space Radio Conference

The fact that the Department of Defense and the Communication Satellite
Corporation were pursuing different systems simultaneously during 1962
and 1963 and that neither institution had finalized plans for primary working
systems had important implications for US radio spectrum planning before
the 1963 ITU conference. To cover all contingencies, the United States proposed very large blocks of frequencies. The planned commercial system being
organized by Comsat was especially useful in providing cover for military
frequency needs. The ITU did not distinguish between military and nonmilitary users of the spectrum. Thus, the United States did not have to explicitly
state that its frequency proposals covered military needs.[60]

As the institution responsible for IRAC, the Office of Civil and Defense
Mobilization (OCDM) took the lead in 1960 in preparing the official US proposal for the 1963 Space Radio Conference. In August, Fred Alexander, the
deputy assistant director for telecommunications, instructed all government
agencies to submit anticipated space communications requirements through
the year 1970.[61] This extended period was chosen to maximize use and to
cover all contingencies. Government officials at the mobilization office discovered that the government needs for space frequencies were "substantial."[62]
Especially important were the formal proposals by the military agencies in
January 1961, which were based on "comprehensive studies" during the previous year.[63] Altogether, the total requests of government agencies added up
to a staggering block of frequencies 25,000 MHz wide. IRAC worked with its
FCC representative to consolidate the total requests of government users and
then submitted a formal report to the FCC in May 1961. After consulting industry, the FCC worked with the OCDM to arrive at a tentative final position.
This government agency and the agency that succeeded it in 1962, the Office
of Emergency Planning, which also had authority over IRAC, thus played
dominant roles in determining the official US position on space frequency
needs.[64]

Key individuals at the mobilization agencies involved in preparing for the 1963 Space Radio Conference included Fred Alexander, who became director of the Telecommunications Office in late 1961; Irvin Stewart, the first director of telecommunications management in the mobilization agency when President Kennedy created the post in February 1962; Ralph L. Clark, an expert appointed to assist Stewart after the Cuban missile crisis; and the longtime executive secretary of IRAC, Paul D. Miles.[65] Clark's background underscores the importance of national security considerations in the development of radio allocation policy in the United States. After earning a bachelor of science degree in electrical engineering at Michigan State University, Clark worked for twelve years as an engineer with the radio and communication regulatory commissions in the Department of Commerce. During World War II, he conducted communications-related research for the navy, first as a lieutenant and later as a commander. He continued to work with defense interests after the war as a civilian. From 1946 to 1949, he served as the director of the Programs Division of the Research and Development Board in Washington, responsible for collecting information about research and development being conducted by military agencies. He then spent six years at the Central Intelligence Agency (CIA) as the deputy assistant director for scientific intelligence; one year as the staff director for the President's Committee on Telecommunications Policy and Organization; two years as the special assistant to the deputy director of the CIA; three years as the manager of the Washington office of the Stanford Research Institute, responsible for coordinating research programs conducted by the institute for government agencies; and four years as the assistant director of defense research and engineering for the DOD Communications and Data Processing Division, which included work with the defense communications satellite programs.[66] In April 1961, Robert Nunn Jr., special assistant to the NASA administrator, described Clark and one of his colleagues, John Rubel, as the two "most knowledgeable people in the Pentagon" on communications satellites.[67] Clark's appointment to the mobilization agency thus reflected the need for someone with a strong national security and intelligence background to help the federal government respond especially to the crisis in electronic communications identified with the Cuban missile crisis and to help coordinate military and civilian planning for satellite communications.

Although officials connected with the mobilization agencies played a dominant role in preparations for the 1963 Space Radio Conference, key officials at the State Department were also important. Especially significant were staff in the Bureau of Economic Affairs during the early 1960s, including G. Griffith Johnson, the assistant secretary; W. Michael Blumenthal, the deputy assistant secretary; Francis Colt de Wolf, chief of the Telecommunications Division; Arthur L. Lebel, who succeeded de Wolf as division chief in 1962; and William Carter, special assistant for international space communications, who had a special interest in overseeing the organization of the global satellite communications system. State Department staff in the Bureau of International Organization Affairs also helped prepare for the 1963 Space Radio Conference. The key individuals in this bureau involved with spectrum policy during the early 1960s were the two leaders, Assistant Secretary Harlan Cleveland and Deputy Assistant Secretary Richard N. Gardner. Finally, coordination between the State Department and the agencies involved with national security was the responsibility of U. Alexis Johnson, deputy undersecretary of state for political affairs. He played an important role preparing for the 1963 Space Radio Conference by also taking account of the needs of the DOD and the intelligence agencies.

In May 1961, the OCDM submitted preliminary recommendations for the 1963 Space Radio Conference to the State Department, which then sent them in the fall to select US embassies for overseas consultation.[68] These views were incorporated into a new proposal completed one year later; further consultations were subsequently conducted with other countries, especially in Europe. As envisioned by the United States, the satellite communications service would require "considerably more spectrum space than all of the other space services."[69]

Because the communications satellite proposal was the biggest request (equivalent to a band of frequencies 2975 MHz wide), it was also the most controversial. In response to resistance from countries interested in taking advantage of the same frequencies for other uses, the United States first extended the period in which all anticipated needs would be considered from 1970 to 1975, and then after further consultations, decided to extend the time to 1980 and increase the frequency proposal for satellite communications to 3000 MHz. In January 1963, the United States emphasized that the estimate of 3000 MHz was "based on the total world traffic requirements up to at least 1980 regardless of the number of systems which may develop ultimately." The proposal would also cover all potential users of satellite communications, including "government, commercial, and private users."[70] And it would take into account satellites using "wide-band and narrow-band channels; in polar, inclined, and equatorial orbits." Wide-band channels would embrace "telephone, telegraph, television relay, facsimile, and data transmission services," including the global distribution of meteorological data. By emphasizing that they were looking out for the world's needs through the next two decades, US government officials believed they would be able to "overcome any uncertainty" that 3000 MHz was "desirable."[71] By setting aside generous blocks of frequencies for satellite communications, they hoped to avoid past problems when allocations for new radio services had "rarely been adequate."[72]

State Department officers and officials with the mobilization agency recognized that to influence the 1963 Space Radio Conference, they first needed to win over Study Group IV of the International Radio Consultative Committee (known by its French acronym, CCIR). The CCIR was established in 1927 to give the ITU technical advice. The United States took the lead in establishing Study Group IV in 1959. Study Group IV had a specific mandate to investigate the frequency needs of space communications systems and to develop technical recommendations for the 1963 conference. The ITU encouraged the CCIR study groups to hold meetings away from the headquarters in Geneva, Switzerland, "to give engineers and experts the opportunity to observe at first hand the technical advances and developments in other countries." State Department officials took advantage of this policy when they invited Study Group IV to meet in Washington, D.C., during spring 1962. Normally, study groups met in the home countries of the chairman or vice chairman. The chairman of Study Group IV was from Italy, and the vice chairman from Switzerland. Charles Bohlen in the State Department's Bureau of International Organization Affairs argued that if the group met in either of these countries, "there would be very little to see regarding developments in space communications." He emphasized that, by contrast, "the United States is the leading country in the development of technical radio facilities and has made more advances in the communication phases of space studies than any other country." To convince other US government officials to agree to sponsor and fund the meeting, Bohlen reminded them that the country would "gain considerable prestige if Study Group IV could hold its first meeting in this country."[73]

All parties agreed to meet in Washington, D.C. The chairman of Study Group IV, Ivo Ransi of the Institute of the Ministry of Communications of Italy, visited Washington prior to the meeting to gain further information about US activities and to begin planning separate committees to study specific problems.[74] The State Department made many of the arrangements for the meeting. To develop an "appropriate program of technical field trips, lectures, and social entertainment for the visiting foreign delegates," the department established an "ad hoc government-industry committee" under the leadership of Andrew Haley, the general counsel for the American Rocket Society. Industry involvement was essential, according to the State Department, to demonstrate to the world "the effectiveness of the United States free enterprise system."[75] Participation by industry would also help pay conference expenses. The department requested all participating companies to pay for "the official and non-official activities for which government funds were not available."[76]

More than two hundred individuals from twenty-eight countries participated in the Washington meeting of Study Group IV in March 1962. These included representatives of thirteen private companies, seven international organizations, six scientific and industrial groups, two other specialized agencies of the United Nations (UN), and three separate groups within the ITU. The United States had the largest national delegation—more than forty individuals. The United Kingdom sent fourteen representatives; the Soviet Union, ten; and France, eight. Of US companies, GTE sent three representatives, AT&T sent five, and RCA sent two.[77]

The official report by the US delegation judged the Washington meeting a success mainly because the other delegates tacitly accepted the US report on "feasibility of sharing between communications satellite relays and terrestrial microwave relays." If the ITU rejected the proposal for sharing, then the 1963 Space Radio Conference would need to find a way to massively reorganize the allocation of a large section of the radio spectrum. The Americans decided it would be difficult—if not impossible—to get member countries to agree to a new arrangement of this magnitude. As long as the radio conference agreed to the recommendation for sharing, then experts could try to develop specific regulations to limit interference, such as stipulating the geographic separation between satellite Earth stations and other radio transmitters on the ground. The United States also succeeded in convincing many of the foreign delegates at the Washington meeting of the CCIR study group, "not so advanced in space communications technology," how space exploration would "be of assistance to them."[78]

A contributing factor in the CCIR's support of the official position of the US government was a change that occurred in the role of US delegates following the 1959 ITU meeting. Traditionally, the US government gave its CCIR delegates the freedom to pursue technical studies independent of government influence. They were expected to pursue "frontier studies" that government officials would adapt as necessary to the national policy goals of the United

States.[79] But starting in 1960, during the period of increased tensions with the Soviet Union following the downing of Gary Powers's U-2, US government officials involved in frequency planning "made arrangements for the regular participation of the Executive Secretary of the IRAC in the meetings of the Executive Committee responsible for US participation in CCIR activities."[80]

Reflecting the global Cold War concerns of the early Kennedy administration, government pressure intensified on American CCIR participants during 1961. In January, Francis Colt de Wolf, chief of the Telecommunications Division at the State Department, agreed to a request from Fred Alexander, the deputy director for telecommunications in the mobilization agency, to have "United States nationals participating in various international telecommunications meetings . . . speak with the same voice in seeking [a] common objective."[81] The State Department agreed in February to first consult with IRAC about US CCIR study group reports "before such reports are approved by the CCIR Executive Committee as a contribution to the US."[82] Thus, because IRAC largely served the Cold War interests of the United States, American delegates to the CCIR also increasingly became tied to the national security state.

Global Campaign for Space Frequencies

To show third world countries that they could benefit from the space race, US officials sought ways to promote technical assistance programs linking ITU member nations to the planned global satellite communications system. Starting in 1961, the United States encouraged the adoption and implementation of a resolution instructing the UN "in consultation with the ITU . . . to give sympathetic consideration to requests from member states" for assistance in surveying "their communication needs . . . so that they may make effective use of space communication."[83] Even before the establishment of Comsat, the United States had recognized the importance of communications for Cold War foreign policy. Following the Gagarin flight in May 1961, the government committee organized by the State Department to make recommendations on issues involving international telecommunications, the Telecommunications Coordinating Committee (TCC), recommended that the federal government initiate "a program of technical assistance directed to the special needs of the less developed countries of the world for more adequate communication facilities." "The extent to which these nations will gravitate toward the influence of the free world," according to the committee, "will be determined largely in terms of their ability to utilize the new technologies of communication so essential to their social, economic, and general well being." This aid might come in different forms, from "education in modern telecommunication theory and practice" to "financial aid for the construction of facilities."[84]

This view of the importance of foreign aid to the Cold War especially reflected the dominant influence of Walt Whitman Rostow on US foreign policy during the early 1960s. Rostow, who served first as the White House national security advisor and then as chairman of the State Department's Policy Planning Council, believed the Cold War would mainly be fought in the third world using development tools and theories. Communications would play an especially important role not only because of the connection to propaganda but especially because Rostow and other theorists in the Kennedy administration believed the promotion of modern communications would play an important role in the process of modernization. To support the battle against communism, according to David Halberstam, Rostow wanted to see "television sets in the thatch huts of the world."[85]

The United States not only used the newly formed Agency for International Development (AID) and other federal programs to provide direct aid to third world countries but also encouraged the ITU and the UN to develop technical assistance programs serving Cold War ends. In some cases, the need to provide technical assistance for communications systems so that the United States and its allies could compete globally with the Soviet Union was immediate. During 1960, the United States encouraged the ITU to increase its technical assistance to the Congolese civil communications system to prevent the Soviets from gaining influence through their use of similar programs. In August the US embassy in the capital, Leopoldville, informed Washington of the poor shape of the country's communications and warned of continued deterioration that would provide the Soviet Union, which had already sent experts to the Congo, with an opportunity "to move in" to support civil telecommunications. "If ITU can get funds to furnish necessary technicians soon," the embassy advised, "the likelihood of Soviet infiltration will be greatly reduced, if not eliminated." By the end of the month, the ITU had authorized technical assistance funds for thirty-nine communications experts. Thirty-two were already in the country.[86]

The Washington meeting of CCIR Study Group IV had provided US government officials with an important opportunity to contact 27 countries that belonged to the ITU; however, after the meeting, they realized they ideally needed to contact nearly all the approximately 120 members, especially less developed countries such as the Congo, to gain support for US space needs at the 1963 Space Radio Conference. C. W. Loeber in the Telecommunications Division of the State Department warned that "unless the allocations adopted at the Radio Conference are supported by essentially all the members of the union there is danger that space communications may be interfered with seriously by many countries which do not accept the allocations adopted by the Conference."[87] During summer 1962, government officials began formulating plans to send small teams of experts to member countries of the ITU, not only to nations active in the CCIR but also to "the remaining country members."[88] Reflecting the dominant role of Cold War concerns, the Office of Emergency Planning initially played a more important role than the State Department in formulating these early plans.[89]

Officials believed this "missionary" work was particularly important in third world countries because many had shown "indifference" to the needs of space communications. The teams of experts would explain the importance of the conference, convince each nation to participate, explain to small countries how to participate, and—most important—"sell" the US proposals. Irvin Stewart, director of telecommunications management in the Office of Emergency Planning, recommended having teams of two or three experts (approximately thirty-five individuals altogether) visit seventy-four different countries.[90]

The United States sent teams of experts from various government agencies to the major countries in Latin America. American communications companies traditionally had played an important role in developing communications systems in the region. Even before the Cuban missile crisis and the establishment of the NCS, US officials were particularly concerned about enhancing the quality of communications networks in the Americas for both economic and national security reasons.[91]

The US government also sent teams of experts to major countries in Asia, but Africa was different. In general, the State Department decided to concentrate attention "in those areas where the greatest good is likely to be produced." Special US teams were not sent to most African countries partly because, according to G. Griffith Johnson, assistant secretary of state for economic affairs, "relatively few" had "experts who could understand the technical and scientific aspects of frequency allocations for space activities." But Johnson did believe in impressing African countries "with the great importance to them of space communications and with the need for representation from their governments at the 1963 EARC." He also decided to defer to the governments of the United Kingdom, France, and Belgium, because "these former metropoles" were "still quite influential in matters of telecommunications in the

African territories which were formerly their colonies." From experience with telecommunications negotiations, the State Department had learned that "more resentment . . . than support for the United States could be expected from visits of United States teams" to British Commonwealth countries and Francophone countries in Africa. This was another reason that the United States focused more efforts on Latin America than on Africa. Two important exceptions in Africa were Liberia and South Africa. Because Liberia was established by ex-slaves from the United States, the relationship between the two countries was close. The United States did not need to send teams to South Africa because as early as August 1962, the country had already informed the United States that it did not have any objections to the preliminary US proposals for space frequencies.[92]

Teams of US experts were especially active overseas beginning in summer and fall 1962. In November, the month following the Cuban missile crisis, experts from different government agencies attended a conference in Bogota, Colombia, discussing the development of a Latin American telecommunications network. During another meeting in Bogota four months later, US representatives set up demonstrations and exhibitions using photographs, slides, and models of satellite components to educate delegates attending the conference about the planned global satellite communications system.[93] Because these Latin American countries also largely did not have experts "with the necessary technical background to make discussions with the United States teams very promising," the United States relied especially on spectacular demonstrations to emphasize the benefits of space communications. During this same period, experts from the United States met with telecommunication officials in some of the smaller European countries, including Spain, Switzerland, Sweden, Norway, and Denmark.[94] In both Europe and Latin America, officials reported that "the governments contacted intend to be represented at the EARC [Space Radio Conference] and that they were sympathetic to the United States proposals."[95] The State Department also instructed its embassies around the world, including missions located in Asian and African countries, to inform appropriate foreign government officials about the frequency proposals the United States planned to submit at the conference and to keep the department informed about their responses.[96]

The United States conducted more formal planning with its major NATO allies in preparation for the radio conference in a series of meetings beginning in summer 1962. A delegation on the way to a NATO meeting in Rome stopped off in London in June to consult with UK civil and military representatives to "reconcile, insofar as possible, conflicting views in the subject . . . in order to avoid unnecessary USA/UK disagreements on the floor of the conference."[97] In November, State Department officials invited communications officials from Canada, the United Kingdom, France, Italy, and Germany to a "six nation" meeting in Washington to discuss frequency allocations for space communications.[98] Two other important meetings were held in Europe during March 1963: a meeting in Paris of the European Radio Frequency Agency and a "seven country" conference in London.[99]

According to US plans, the global communications satellite service would require "considerably more spectrum space than all of the other space services"; however, these other general uses of communications for space needs, including military uses, had to be considered in negotiations with other countries, especially the Soviet Union.[100] The only information the Americans had about Soviet views derived from discussions at the Washington meeting of the CCIR study group in March 1962. The State Department rejected Soviet proposals for space frequency allocations presented at this conference "because of their inadequacy and their conflicts with important military frequency needs." While the United States had requested the equivalent of a block of frequencies 3000 MHz wide, the Soviet Union had proposed only 900 MHz. Officials suspected that the Soviets had made this proposal "in full knowledge of these conflicts." Griffith Johnson cautioned Edward Bolster, director of the Department of Transport and Communications in the Bureau of Economic Affairs of the Department of State, against giving any indication about which frequencies the military required. If the Soviets asked "whether the United States expects to utilize a global satellite communication system for the movement of its military traffic," he instructed Bolster to "indicate that the United States regards its military traffic as a form of government traffic. As such, it will be passed over all available communication modes, including cables, radio, and satellite relay systems."[101] During the discussions with the Soviet Union prior to the Space Radio Conference, the Soviets continued to make the same requests for space frequency allocations that they had made during the Washington meeting of CCIR Study Group IV.[102]

In general the Soviet Union had less need than the United States for space radio frequencies, but it is important to realize more specifically that the Soviets were less interested in global satellite communications because the country had only about 5 percent of global communications traffic.[103] And although the Soviets could have used communications satellites to serve remote areas of the vast country, the government was already committed to completing an extensive terrestrial microwave-relay system that would provide television coverage and other forms of electronic communications to most of the country.[104] The Soviet Union did build a limited communications satellite system called Intersputnik during the 1970s, but because it mainly served countries in the Soviet bloc, its global impact was limited.

Negotiating Technical Boundaries

Especially when US representatives negotiated with the Soviet Union over space frequency allocations, they tended to emphasize that the ITU and its subsidiary organizations primarily dealt with technical or scientific issues divorced from social, economic, or political factors. When the Soviets raised a controversial issue, US negotiators complained that they were injecting "a political question into our technical proceedings."[105] This form of boundary work proved to be a useful strategy for managing conference proceedings. When a proposal was circulated one month before the Space Radio Conference dictating that any country launching a communications satellite had to make such facilities available to other countries "on an equitable basis," Richard Gardner, deputy assistant secretary of state for international organization affairs, complained that "the injection of these major political and institutional questions could not fail to provoke serious controversy and prejudice the successful conclusion of the technical business of the conference." He claimed issues such as these were inappropriate because the members had reached an agreement that the ITU would confine the conference proceedings "solely to technical questions relating to frequency allocation."[106] One US official encouraged representatives attending CCIR and ITU meetings to provide "thoroughly engineered reasons" for requesting space frequencies.[107]

This strategy of trying to limit evaluations of frequency allocations to technical issues helped the United States contain the involvement of communist countries in the activities of the ITU. The State Department instructed the US representative to the Washington meeting of CCIR Study Group IV to oppose "any attempt to seat a delegation from Communist China in any capacity and/or to invite representatives from Outer Mongolia, North Korea, North Vietnam or East Germany to attend the meeting." He was to oppose their participation "on the grounds that it is completely out of order for a technical group such as this to involve itself with such questions."[108]

But communist countries also used this strategy of technical legitimation to avoid political controversy. Soviet and Cuban delegates to a meeting in Colombia of an ITU organization involved in planning improvements for tele-communications in Latin America argued that their mission was purely a "technical one." The US embassy in Bogota reported to Washington that "the Soviets shied away from making predictions or discussing political subjects . . . They were technicians and would not say whether or not they were Communist Party members."[109]

Despite the attempt by participants in ITU meetings to limit proceedings to technical issues, both sides recognized privately that the technical problem of determining frequency allocations inevitably involved political considerations. The assistant secretary of state for international organization affairs, Harlan Cleveland, acknowledged that ITU staff members should have the highest scientific or professional credentials, but he also emphasized the "desirability of assuring that more United States nationals are employed by international organizations."[110] As we have seen, during the early 1960s, US officials did not expect technical experts to set aside national interests when evaluating frequency allocations. In July 1963, an AT&T employee wrote a State Department official involved in organizing the US delegation to a future meeting of the CCIR that "we strongly support your desire to establish a delegation which will be able to protect and advance United States interests in this important field."[111]

The United States attempted to directly influence the appointment of ITU officials. The State Department recognized especially the importance of a US citizen serving as secretary general of the organization. The fact that this post was held by an American during the period leading up to the Space Radio Conference undoubtedly strengthened the US position. The congratulatory letter sent by Assistant Secretary of State Francis Wilcox to Gerald Gross after ITU members elected him to head the organization in January 1960 stressed that "it is heartening to know that we have an American at the helm."[112] When it seemed that the leadership role would become vacant during summer 1963, the State Department lobbied other countries to support the appointment of another US citizen, John H. Gayer. US officials argued that "if possible the US should continue to hold the post in view of our vital concern in ITU work on communications developments around the world, particularly in space communications."[113]

The tension between wanting to treat the ITU as a technical organization using professional experts to objectively evaluate scientific problems and recognizing that the organization was inextricably bound up with geopolitical considerations was especially clear in the hiring of staff. For example, in May

1963, a Czech engineer, Miroslav Joachim, became the leading contender for the post of assistant to the director in the CCIR Secretariat. But the United States believed Joachim had been using his position as head of the ITU Staff Association, which was in charge of advising top officials about hiring employees, as a means for the "Soviet bloc to infiltrate the ITU staff." A classified report by the US representative of the Administrative Council of the ITU warned that under Joachim's leadership, "the Association had become a force to be reckoned with by the Administrative Council, the Secretary General, and the other elected officials in charge of specific activities of the Union."[114]

Although State Department officials believed Joachim was an "avowed Communist," they admitted that his professional qualifications were "excellent." He was a PhD engineer who had studied at MIT and had "extensive experience as a representative of Czechoslovakia at conferences and as a member of the staff of the ITU." "Refusal to give him the appointment," the Department of State agreed privately, "would be most awkward, unless other equally well qualified applicants appear."[115] The United States thus worked with other Western governments to find another well-qualified candidate who did not have a "demonstrated bias towards Soviet Bloc interests."[116] Having argued that the ITU should decide such issues as the allocation of space frequencies solely on technical considerations to contain the influence of communist countries in the ITU, the United States could not easily disregard "technical" or "scientific" qualifications when trying to influence the hiring of ITU personnel.

The political nature of ITU decision making not only was an important factor in the way the United States dealt with the CCIR, but also was apparent when US officials interacted with other groups of scientists and engineers advising the ITU, such as a panel of experts organized to study "ways and means" of relieving congestion in the use of the band of frequencies from 4.0 to 27.4 MHz. The chief of the Telecommunications Division of the Department of State, Francis Colt de Wolf, recognized that the members of the panel would not "be representing any particular country or administration." They would make judgments based on their "personal capacities" as technical experts.[117] Nonetheless, officials successfully worked to appoint a US national to the panel, Paul D. Miles of the OCDM, with the understanding that he would take into account not only narrow technical considerations but also "the economics of telecommunications systems" and the needs of "military communications."[118]

Whereas both this panel and the CCIR committees had a formal relationship with the ITU, another important advisory body, the International Scientific Radio Union (URSI), was formally independent of the ITU and all other official government bodies; however, in the context of the Cold War, this independence was by no means definitive. Francis Colt de Wolf acknowledged that "as a non-government group," the members were "not . . . required to seek Government approval of the technical views which they express." At the same time, he also emphasized that without government or "intergovernmental" support, the recommendations of URSI and other nongovernment organizations would lack authority internationally. The State Department applied subtle pressure to convince these groups to consider US interests by stressing how their position would be strengthened "if they could accompany" their views "with an indication that they are not only believed to be technically sound but also are supported administratively by responsible government bodies."[119]

But somewhat inconsistently, officials also valued the independent role of these groups. They recognized that in some ways, the scientific authority these organizations could wield might be just as powerful as the political authority resulting from a governmental connection. This became particularly obvious early in 1963 when officials revised their earlier view that had played down the need for exclusive frequencies for space communications. They now believed that they had to convince other ITU members of the importance of setting aside exclusive bands of frequencies for space communications, especially in connection with scientific research but also for satellite communications. One reason for the new view was the realization that the large geographic separations required for sharing with terrestrial fixed and mobile services would be beyond the extent of most countries. The main reason for the change, however, was the realization following the Cuban missile crisis of the importance of mobile satellite Earth stations for the US military and the NCS. Exclusive frequencies were especially important for these planned mobile Earth stations. When officials realized the importance of exclusive frequencies, they also recognized that scientists and engineers would necessarily play a crucial role in convincing other countries to agree by providing powerful technical justifications.

Nearly one-half of the countries contacted by the United States before the Space Radio Conference insisted "upon some form of sharing with the fixed and mobile services." Several large countries wanted sharing to be "on an equal basis." Irvin Stewart, the director of telecommunications management at the Office of Emergency Planning, convinced the National Academy of Sciences to assist in the effort to convince other countries to adopt exclusive allocations for space communications by emphasizing that it was a technical necessity based on "engineering facts."[120] The State Department asked US scientists to contact "appropriate officials of any scientific research organizations" in other countries and "endeavor to persuade them to try to convince the telecommunication authorities (whose responsibility it is to negotiate for frequency allocations) that exclusive use of frequency bands allocated to space research is essential." Officials stressed a connection between frequency needs for radio astronomy and for general space communications to get scientists to lobby for exclusive allocations. They argued on technical grounds that sharing with terrestrial radio sources would not in all cases be possible for either use of space frequencies. In the case of radio astronomy or other uses of radio frequencies for space research, interference would make accurate scientific observations impossible. But interference could also create safety concerns for both manned and unmanned missions and could disrupt a global satellite communications system.[121]

Irrespective of the problems related to obtaining exclusive bands, convincing other countries to agree to closely circumscribed sharing with space communications was a problem in itself. All the CCIR study groups met before the Space Radio Conference in a general meeting in Geneva (the Tenth Plenary Assembly) from January 16 to February 15, 1963, to discuss, among other issues, the stringent technical requirements that would need to be implemented to prevent terrestrial radio transmitters from interfering with satellite Earth stations operating on the same frequencies. They reached agreement on "basic sharing criteria," including such issues as "the limitations on the radiated power from line-of-sight radio relay systems to avoid serious interference to the proposed satellite systems" and "minimum separations . . . between Earth stations of the communication satellite service and stations of the line-of-sight relay services." The final conclusions of this conference, according to the official report of the US delegation, "were compatible with the interests of this country."[122]

1963 Space Radio Conference

Despite the pleasure expressed by US officials in the results of the CCIR and the preliminary "missionary" work before the opening of the Space Radio Conference in Geneva, during the meeting—which lasted from October 7 to November 8, 1963—several countries opposed US frequency allocation proposals for space communications, not only the Soviet Union but also other nations. As late as the beginning of the third week of the conference, the chairman of the US delegation—Joseph H. McConnell, the president of Reynolds Metals, who was given the title of US ambassador when he was appointed to head the delegation—wrote in frustration to the secretary of state that "although the Conference is half-way through the allotted five weeks, there is no evidence that the USSR, or many of the smaller countries, will accept our frequency proposals."[123] The Soviets, together with some of the less developed countries, initially argued that the meeting "should be provisional pending a future planning conference," but the United States eventually managed to convince the Soviet Union and other countries to compromise.[124] A classified report on the conference by US delegates argued that "in a situation wherein the latent suspicion of the Conference body was directed against the United States, it would have required little impetus to produce a result adverse to US policy and productive of ruptures that would be far-reaching."[125] The Soviet Union could have undermined US efforts by championing the cause of developing nations who feared "a spectrum grab" at their expense.[126] But, as the classified report related, the Soviet Union "had sufficient to gain from the outcome of the Conference that it refused to avail itself of an obvious opportunity to create a breach between the US and the developing nations, or to render the Conference a nullity."[127]

Although a detailed analysis is necessary to understand US-Soviet relations at the conference, in general terms, it is especially important to understand that the conference took place during a period of détente following the Cuban missile crisis in October 1962. After taking the two countries to the brink of war, Khrushchev abandoned his policy of brinkmanship to seek better relations with the United States.[128] The spring of 1963 proved to be a "turning point in the Cold War."[129] Facing new evidence of economic difficulties, Khrushchev was desperate for a disarmament agreement with the United States to halt the escalating costs of the Cold War.[130] By April he was willing to consider a compromise on the status of Berlin and a less than complete ban on nuclear tests.[131] The two sides agreed on concrete actions after Kennedy gave a highly conciliatory speech on the topic of world peace in June. Separate efforts by the German government to compromise with Khrushchev, and the breakdown in relations between China and the Soviet Union, helped pave the way for a relaxation of Cold War tensions.[132] The United States and the Soviet Union negotiated a ban on aboveground nuclear tests in July.[133] They also signed an agreement establishing a "hot-line" linking the respective capitals for urgent communications through telex and radio links.[134] Even before

Kennedy's June speech, NASA and the Soviet Academy of Sciences agreed on a memorandum of understanding establishing cooperative space programs, including experiments with communication and meteorological satellites.[135] This tentative agreement was based on a series of talks at the UN between Professor Anatoli Blagonravov of the Soviet Academy and NASA Deputy Administrator Hugh L. Dryden.[136] Khrushchev's son recalled that his father had been in an extremely optimistic mood regarding US-USSR relations during the period before the ITU Space Radio Conference, especially during late summer 1963.[137]

In addition to McConnell, the main US delegation to the Space Radio Conference included two vice chairmen, twenty-three technical advisers, and four members of Congress serving as official delegates. The technical advisers overwhelmingly represented different government agencies. Three were from the FCC. Four government organizations each sent two representatives—the Department of State, the Department of Commerce, the National Aeronautics and Space Administration, and the Office of Emergency Planning. The army, the air force, the navy, the Defense Communications Agency, and the Federal Aviation Administration each had one employee on the delegation. Only three of the advisers represented commercial companies involved in space research (one each from AT&T, ITT, and RCA); two advisers represented Comsat, and one was a university professor. If more representatives of commercial interests had been included on the delegation, the Soviets could have used this to support propaganda warning that the new satellite corporation would be dominated by commercial interests, especially AT&T. According to the official report of the qualifications of individual members of the delegation, the AT&T employee was included not to represent the company but to "make an important technical contribution."[138]

The importance of satellite communications to the conference was clear in the proceedings and organization. The United States helped arrange to televise the opening sessions across the Atlantic, using the communication satellite *Telstar 2*, launched on May 7. On October 9, U. Thant, the UN secretary general, exchanged greetings from New York City with the ITU secretary general in Geneva. In extended remarks, U. Thant spoke of his hope that through this new development, "man can acquire a global perspective and that this will help him not only in his struggle to achieve a better life on earth, but also in achieving a greater unity of purpose and direction for all mankind."[139] The geosynchronous satellite *Syncom II*, built by Hughes under a NASA contract and launched in July 1963, was also made available to delegates in Geneva to communicate with various people located at the UN building in New York and at NASA headquarters in Washington, D.C. During the last evening of this demonstration, the United States invited members of the press in Washington to interview McConnell and other US officials using the Syncom satellite.[140] US officials based this attempt to gain the support of conference participants using spectacular demonstrations on similar practices during the lead-up to the conference. Fred Alexander, the head of the Telecommunications Division in the Office of Emergency Planning, had urged such demonstrations at the Space Radio Conference, arguing, "It was the experience of the US Teams of Experts that simple exhibits of pictures, slides and film aroused considerable interest and increased understanding of space radiocommunication and its potential."[141]

To ensure that satellite communications received special treatment at the conference, the US delegation was divided into two groups. The first, known informally as the Satellite Communications Policy Group, worked to educate other ITU members about the planned global communications satellite system. The second group included all remaining delegates in charge of the "substantive work of the conference by advocating adoption of US positions, defending such positions when required, and effecting necessary compromises."[142]

During most of the Space Radio Conference, the US delegation was unsure about Soviet motives for initially refusing to agree to US frequency proposals. Because they knew little about Soviet space needs and capabilities, they had to speculate based on a limited amount of information. Recently declassified documents indicate that, in 1962, US intelligence agencies concluded that the Soviets were not working on their own global system or "any other satellite communications system."[143] The Soviet delegation showed little interest in accepting an offer to participate in the global system that the United States was organizing. Based especially on statements from the Soviets about the need for further experiments, US delegates assumed that they were "not as advanced as we are in satellite communication, and would like to delay until they can catch up." McConnell complained to the secretary of state about the Soviets using "various ploys" to delay US efforts, including calling for a future ITU conference on space frequency allocations and treating any recommendations resulting from the 1963 conference as provisional.[144] US officials insisted that "the use of the frequencies had to be sufficiently definitive to permit long-range planning and major investments in a global communications satellite system."[145]

During most of the conference, the Soviets continued to call for a block

of frequencies for satellite communications much smaller than the US proposal—1600 MHz as opposed to 2725 MHz. Especially troubling from the US
perspective was that only about 800 MHz was common to both proposals. As
we have seen, US officials based their large proposal on the "best estimates of
what would be required to handle satellite traffic through 1980."[146] Although
this estimate also took into account DOD planning for a separate communications satellite system, officials with the State Department, the mobilization
agency, and the National Aeronautics and Space Council pressured the DOD
immediately before and during the conference to avoid publicizing the connection.[147] Before the conference, military officials made sure that their discussions with NATO allies about specific frequencies planned for the military
system would not become public.[148]

A specific event involving the military three days into the conference also
raised concerns about the danger of acknowledging the military use of space
frequencies. On October 10, McConnell sent an urgent telegram to the Department of State in Washington warning of the consequences of the "immi-
nent public announcement on the letting of contracts for a military space
communications system." Assistant Secretary of State Griffith Johnson responded by warning the secretary of defense, "Any public announcement
during the course of the Conference which would convey the impression that
the United States is giving priority to military space communication systems,
is likely to generate a political debate on military usage of radio spectrum and
would weaken seriously the chances of the United States obtaining agreement
to United States proposals."[149] Any acknowledgment that the United States
planned to use space frequencies for military communications would have
jeopardized the support of nonaligned countries that were willing to support
space frequencies because of the promised connection to the civilian global
satellite communications system.

An especially important disagreement between the United States and the
Soviet Union involved the issue of exclusive frequency allocations for both
space research and general space communications.[150] The Soviets opposed all
talk of exclusive allocations "as a matter of principle." The United States and
most other countries in Western Europe favored setting aside exclusive bands
for different uses of space radio frequencies. Soviet opposition to a US-sponsored
proposal calling for 100 MHz of exclusive spectrum for communications satellite service in the bands located at 7250–7300 MHz and 7975–8025 MHZ
(for "mobile and transportable earth terminals") contributed to a major im-
passe at the Space Radio Conference.[151] By the beginning of the third week of

the conference, McConnell realized that if the United States was "going to reach an accommodation" with the Soviet Union, there would "have to be some give on both sides."[152]

The Soviets specifically opposed the US proposal for exclusive frequencies because they understood this was particularly important to the US military and the NCS. A separate military satellite communications system using mobile stations in remote areas of the world would need exclusive frequencies to avoid interference from local transmitters.[153] When Joseph McConnell was interviewed in 1985 about his work at the 1963 Space Radio Conference, he argued that US military representatives on his delegation had caused him more difficulty than the Soviets. According to McConnell, the Soviets were "difficult on occasion, but not really, they never did go back on anything."[154] McConnell relied on Joseph Charyk, the recently appointed head of Comsat, for advice about dealing with military demands for space frequencies. McConnell specifically requested Charyk's attendance during the final crucial week of the conference.[155] Charyk had the necessary expertise because of his background as the undersecretary of the air force, responsible for the DOD's satellite programs.[156] According to McConnell, Charyk "was wonderfully helpful because he knew so much about what the military was demanding."[157] During the conference, McConnell had felt that the army's demands were excessive, but he did not have the technical expertise to understand which demands "were valid."[158] Charyk advised McConnell "confidentially" about how the military "was over grasping." McConnell boasted that he then "cut them off at the knees."[159]

Undersecretary of State George Ball had chosen McConnell in response to pressure from several sources to get "some pretty high powered talent on the delegation."[160] Key members of Congress, especially Senator Pastore, chairman of the Communications Subcommittee of the Committee on Commerce, had been particularly critical of the State Department for relying too heavily on technical experts from the Telecommunications Division during previous international conferences, in particular the longtime chief, Francis Colt de Wolf. "By downgrading" policy "in our own State Department" in this way, Pastore had worried that the government was not "giving sufficient importance in the way of prestige for the purpose of formulating policy that will be respected and recognized abroad." During committee hearings in 1961, Pastore had insisted that the head of the delegation to the 1963 Space Radio Conference should be a high-power personality with prestige, importance, and political clout, preferably with the title of ambassador.[161] Before McConnell's appointment to head Reynolds, he had served as president of the National Broadcasting Company and had extensive experience as a tough Wall Street lawyer.[162] McConnell's close work with Charyk at the conference led to his appointment as the chairman of Comsat. A former Comsat employee who worked for McConnell recalled how he intimidated everyone because he was "so strong and dynamic and so forceful and tough and frightening . . . He took people apart."[163]

Despite the "missionary work" undertaken before the conference to educate countries about the benefits of the planned global satellite communications system, several third world countries opposed US efforts to set aside exclusive space frequencies or allow a large amount of sharing. Complicating US efforts was the fact that most countries from sub-Saharan Africa and some of the countries in Latin America failed to attend the conference. The US delegation worked to educate delegates from less developed countries in attendance "in every way possible" but recognized that little could be accomplished "in such a short time." A major issue not foreseen by the United States during preconference planning was that "many of the delegates" attending the meeting did not have "all the authority that we would wish." McConnell wrote George Ball and Assistant Secretary of State Harlan Cleveland that "if we had known all this three months ago perhaps we could have educated them to better effect, but of course no one could have known." McConnell also informed the State Department that multiple small developing countries were "fearful that if our proposals are adopted, we will usurp so much of the spectrum that they will be foreclosed from taking advantage of it later." In a revealing passage, he pointed out that "the history of frequency allocations, up to now, gives them no cause for any contrary conclusion."[164]

McConnell believed that the United States could get a majority vote at the Space Radio Conference even without the support of a number of developing countries (or even without the support of the Soviet Union), but he feared that this "might not be a real victory" because countries voting against allocation proposals still had the right to count themselves out of specific requirements through the use of "country footnotes" to frequency allocation tables.[165] US delegates worried especially about Cuba exempting itself from provisions of international allocations. "With Cuba only 90 miles from the United States," one official warned, "Cuban footnotes could adversely affect present and future US operations such as the Atlantic missile range and radio astronomy."

The use of footnotes would also set a bad precedent, especially for the Western Hemisphere. No country in this region had ever requested footnotes for exemption from international allocation tables.[166]

Multiple countries offered proposals or resolutions at the conference aimed at preventing "presently space-oriented countries" from gaining exclusive use of large blocks of the radio spectrum, but the United States managed to control major opposition to US allocation proposals. The classified report by US delegates reported that an "undercurrent of suspicion . . . surfaced in either speeches or resolutions by Israel, Morocco, Algeria, Cuba, Kuwait, United Arab Emirates, Ethiopia, Pakistan, the Soviet bloc, and others."[167] American representatives were particularly baffled by Israel's important role in stirring up opposition to setting aside large blocks of exclusive frequencies for space communications. They complained that the resolutions Israel offered would "cause major difficulties and delay" by vitiating allocation agreements and undermining much of the work already undertaken by the United States and its allies.[168] The representatives from Israel at the conference gave US delegates the impression that the country was trying "to be the spokesman for the black African countries in defending the interests of the under-developed countries who must receive their fair share of frequencies and communication channels."[169] A telegram to the State Department argued that Israel was "playing politics [with] African countries to appear [the] champion of small powers."[170]

The State Department enlisted the assistance of its embassy in Tel Aviv to pressure the Israelis to modify or withdraw the major resolution that opposed US space policy.[171] The resolution complained that space radio communication was being treated as "both the privilege and the exclusive possibility of great countries only." It requested that the Space Radio Conference "abandon or at least modify the present practice of first come first served" and establish "some form of Space Communication Administration . . . entrusted with the responsibility for insuring the global interests . . . of all member states" of the ITU.[172]

The comments, early in 1963, of the director of communications systems at NASA may have provided third world countries with a reason for suspecting US intentions. The director questioned to what extent "less developed" countries would participate in a global system. "All nations," he believed, "will not benefit equally from participation in a world-wide communications satellite system; indeed, some nations perhaps should not participate at all. Clearly, only a small number of countries should have satellite ground stations."[173] After they received assurances from the United States that its proposed global satellite communications system would benefit developing countries, the Israelis agreed to US demands. A US delegate in Geneva reported to the State Department that the "Israelis appeared surprised and reassured to learn that [the] US envisaged [a] single, global commercial system."[174] The American delegates at the conference helped the Israelis redraft the key resolution "so as to be acceptable to [the] US and at [the] same time offer them means to save face and maintain influence with the developing countries to which they had made commitments—including Ghana, Ethiopia, Liberia, and Iran."[175]

US efforts also faced difficulties when the International Frequency Registration Board (IFRB)—a permanent part of the ITU responsible for keeping track of the international use of the radio spectrum and for advising countries about actual or potential interference—supported developing countries and the Soviet Union in their attempt to delay the establishment of a permanent allocation for space communications and research. The United States organized a vote of eighteen to four to defeat the IFRB resolution in one of the seven committees organized at the Space Radio Conference. The four votes in favor came from delegates representing the Soviet Union and three countries from its Eastern European bloc. A compromise satisfying the Soviets was reached through the introduction of vague language in a recommendation calling on the Administrative Council of the ITU to continuously evaluate whether conditions warranted the convening of another conference.[176] A similar method was used to convince the Soviets to modify another proposal requiring that all ITU members coordinate their space activities with any other member also planning to use space radio communication. US officials considered this unacceptable because they worried that it might mean that any country would be able to "effectively block" another country's plans for the use of space.[177]

Compromise and Closure

All the major countries with an immediate interest in space exploration had to make concessions to reach a final agreement on frequency allocations when they realized this was necessary during the third week of the conference. The heads of the delegations representing the United States, the Soviet Union, and France held separate talks during the fourth week of the conference to reach agreement on the most controversial negotiations of the meeting, related to the allocation of different blocks of space frequencies. Although the final negotiations involved the allocation of relatively small blocks of frequencies throughout a major portion of the higher frequencies in the radio spectrum for such services as tracking, telemetry, space research, meteorological satellites, and navigational satellites, the main focus of the final negotiations was the allocation of large blocks of frequencies for satellite communications.[178] In a "joint compromise proposal," the Soviets agreed to allow the use of satellite communications in a larger band of shared frequencies than they initially believed was necessary. The United States and the other Western countries agreed to "reduce the amount of readily useful spectrum space in their proposals from 2725 MHz or more to 2000 MHz" and to relax some of the technical requirements for sharing. The final agreement authorized 2800 MHz for satellite communications, but only 2000 MHz of this included bands originally requested by the United States. This block of 2000 MHz was divided into four bands, each 500 MHz wide. Two of these bands (3700 MHz to 4200 MHz and 7250 MHz to 7750 MHz) were for communication uplinks (satellite to Earth); the other two (5925 MHz to 6425 MHz and 7900 MHz to 8400 MHz) were for downlinks.[179]

The joint talks did not result in agreement about whether to allow exclusive spectrum space for satellite communications. The three countries authorized one of the major committees of the conference to vote on this issue. A majority of the members of Committee 5, which dealt with decisions about allocations, decided through a secret ballot to "accept" the US proposal for 100 MHz of exclusive frequencies, but this did not represent complete acceptance because they also decided to allow existing operations to continue until January 1, 1969.[180] McConnell was then able to get the military representatives on his delegation to agree to accept the limitations on these exclusive frequencies, which they planned to use for mobile satellite Earth stations. Because of the efforts by the United States to avoid publicizing potential military frequencies, most representatives at the conference, especially from smaller countries outside Europe, would not have understood this connection.

McConnell's official report argued that "the overall objectives of the United States were approved by the Conference which adopted the majority of the US proposals in substance." He expressed confidence that "US programs in the various areas of space radiocommunication," especially satellite communications, "could proceed satisfactorily." They would not have to worry about interference in the bands shared with terrestrial radio sources, according to McConnell, because "the Conference largely succeeded in accomplishing the difficult task of superimposing the allocations for the communication satellite service on those already made to terrestrial fixed and mobile services by prescribing technical criteria essential to the avoidance of mutually harmful interference." The official report specifically praised the extensive preparatory work undertaken well before the opening of the Space Radio Conference. Although several countries continued to suspect US promises to develop a system benefiting all nations, the preconference coordination did result in crucial support during the conference of "at least 19 European countries," as well as Japan, Canada, and several Latin American nations. McConnell strongly urged the Department of State to support "such pre-Conference coordination" with future international telecommunications meetings.[181]

Despite McConnell's judgment that the conference successfully met US interests in space, his delegation also had to accept compromise to reach a useful agreement. The Americans had to agree to reduce their request for the amount of "readily useful spectrum space" for space radio communications, and sharing criteria with existing terrestrial services would not be as rigorous as the United States had originally proposed. And of the relatively large number of regions in the spectrum the United States wanted to keep strictly exclusive for use by space services, the conference accepted—with absolutely no exemptions—only two narrow bands (one for radio astronomy and the other for radio-navigation satellites). The other requests were essentially rejected, including—as we have seen—the proposal for 100 MHz of completely exclusive frequencies for satellite communications.[182]

The Kennedy administration was open to compromise during this period because it was serious about improving relations with the Soviet Union. During late fall 1963, this objective was especially important because Kennedy was under political pressure to find ways to reduce the escalating costs of the Apollo moon program.[183] He was willing to consider the possibility of cooperating with the Soviet Union in manned space operations, including a joint moon program. Soviet representatives first broached the possibility informally.[184] President Kennedy then formally proposed the idea in a dramatic speech at the UN General Assembly on September 20. On November 12, he directed the NASA administrator, James Webb, to assume responsibility for investigating the feasibility of a joint moon program or other substantial cooperative space projects.[185] Both countries were interested in considering cooperative space projects partly as a way to reduce costs but also because of the political advantage each nation would gain by demonstrating leadership to the world in the cooperative and peaceful exploration of space.[186] Kennedy's death on November 22 halted these plans, but the fact that they were taken seriously at the time underscores the general climate of cooperation and compromise during the period of the Space Radio Conference.[187]

One result of US willingness to compromise was that some countries did decide to include footnote exemptions in the international frequency tables. McConnell was particularly unhappy because for the "first time in the history of international radio regulation," a delegation from a country in the Western Hemisphere, Cuba, "deviated from the radio frequency allocations agreed to by all other ... delegates" in the region. Cuba indicated it would use footnotes exempting participation in most of the agreements involving space frequencies, including the use of specific bands for satellite communications. Because of worries that Cuban radio sources might interfere with important US space and radio operations in the Caribbean, the United States declined to honor Cuban footnotes. McConnell believed this represented the "first time" the United States had to take such a position "on decisions of any world-wide international conference."[188]

For the specific case of frequencies allocated almost "exclusively" for satellite communications, the twenty countries allowed to continue operations until 1969 added footnotes indicating that "in their countries the fixed and mobile services would continue to have primary status, sharing the bands 7250–7300 and 7975–8025 MHz coequally with the communication-satellite service." This action underscored the fact that many developing countries still were not convinced that they would benefit from the planned global system, especially in comparison to terrestrial services using the same frequencies.[189] In achieving closure, the United States had conducted most of its high-level negotiations at the conference with the Soviet Union. The classified report by US delegates pointed out that "in accomplishing this necessary result the delegations were unable to allay the fears of the developing countries that their rights and future interests were being jeopardized." "Unfortunately," the report pointed out, "the time and tempo of a conference militates against the intimate contacts necessary to convert the suspicions of the smaller and developing countries into more than a reluctant acquiescence."[190]

Despite the shortcomings of the allocation proposal, in general the concessions made by the Soviets were more substantial than the compromises made by the United States. Officials acknowledged that the total authorization of 2800 MHz for satellite communications could theoretically handle eight to nine thousand telephone circuits and many television channels. Since the government's official projected estimate for 1980 of the total number of voice channels required "to and from the United States to Latin America and to Europe, the Near East, and Africa" and for "all other telecommunications requirements" was approximately 13,500, the final spectrum authorization of the Space Radio Conference would provide adequate service for many years. Cables and conventional radio circuits would continue to accommodate some of this future demand, but the government forecast that a satellite system using 2800 MHz would be capable of satisfying approximately two-thirds of this predicted need.[191]

The United States also hoped that lingering suspicions about the need for special space radio frequencies would disappear when ITU countries became involved in the global satellite communications system. Developing countries, in particular, would get a better sense of how they would benefit from the new technology and would be more likely to accept stricter regulation of bands used by the new service to prevent interference. The major motivation for the United States was to use the satellite system as part of a Cold War struggle to strengthen ties with other regions of the world, especially countries in the Global South. But the new satellite system would also benefit the United States economically by linking the country to new markets around the globe, and the largely American-controlled satellite system would allow former European colonies to have not only political independence but also communications independence as they could bypass European-controlled undersea cables and other colonial networks.

The 1963 Space Radio Conference played an especially important role in helping the United States solidify alliances with Europeans and other allies while also providing valuable lessons for dealing with less developed countries. Although the willingness of the Soviets to compromise partly reflected a new period of détente in 1963 following the Cuban missile crisis, the ability of the United States to convince the Soviet Union to compromise at the 1963 Space Radio Conference was also made possible by the specific strategies employed by the United States and key European allies at the conference. Particularly important in the development of spectrum policy was the strategy involving boundary work. Technical issues and technical experts were central to the work of the ITU, but because the organization was intergovernmental, political factors always had to be considered.

Although the United States was not entirely successful in gaining the support of less developed countries, US diplomats learned from the experience that they needed to follow up with more extensive "missionary" activity in Africa and other regions. The 1963 conference also underscored the importance of technical assistance to and technical education of poorer countries.

The fact that many of these countries did not have experts who understood the technical issues involved in spectrum policy hampered diplomatic efforts. Spectacular demonstrations helped demonstrate the broader significance of space exploration and research, but specific decisions involving frequency allocations demanded advanced technical training. The efforts to develop international frequency allocation policies for space communications during the late 1950s and early 1960s reflected the beginnings of a new era in the history of the ITU. No longer would the East-West conflict dominate conferences. During this new era of global Cold War, the United States also had to deal with the growing importance of nonaligned countries and a new North-South conflict.

My analysis in this chapter of the initial attempts to allocate frequencies for a new radio communication "service" is especially important for providing a deeper understanding of the relationship between national security and communications policy during this crucial period of the Cold War. Although the radio spectrum has traditionally been treated as a common resource, it has also been the focus of intense conflict. US communications policy became increasingly driven by national security concerns during the 1950s and early 1960s, especially during the Cuban crises in 1961 and 1962. Military and mobilization agencies played a crucial role in determining the official US position on space frequency needs for the 1963 ITU Space Radio Conference. The Cold War radio spectrum reflected a fundamental tension between cooperation and conflict. The United States and the Soviet Union competed on a global scale not only for military dominance but also to win over hearts and minds by demonstrating both materially and symbolically which country had the superior system and which country was more committed to world peace and understanding. The US government viewed communications policy in general and the planned global satellite system in particular as a key aspect of a soft-power initiative to prevent developing countries from going communist.

Despite the link between US spectrum proposals and military needs, this connection could not be publicized because it would alienate the many nonaligned and poorer countries whose support was desired. To convince these countries to agree to set aside valuable frequencies for space, the United States had to convince them that they could also benefit from space exploration and research. This was a major motivation for the establishment by the US government of the first global satellite communications system. By emphasizing that all countries were eligible to join, the global system would not only demonstrate the practical side of space technology but also win over hearts and minds to the US position by showing in a spectacular way the country's superior ability to promote science and technology. Although the 1963 Space Radio Conference set aside several blocks of frequencies for many different uses for space communications, satellite communications played a central role at the conference. The commercial global system was especially useful in justifying the need to set aside large blocks of frequencies for all uses of space, including a separate satellite communications system being developed by the US military. Thus, for the United States, management of the Cold War radio spectrum involved national security considerations based not only on the needs of global military preparedness but also on the need to wage a symbolic and material struggle for hearts and minds around the world.

CHAPTER FOUR

Organizing the First Global Satellite Communications System

Thirteen nations signed an "Agreement Establishing Interim Arrangements for a Global Commercial Communications Satellite System" on July 24, 1964.[1] Other countries were then invited to add their signatures on August 20, and Intelsat was officially established. Although the major objective of the United States was to organize a global system that would include poorer countries from the Global South, US officials made an important decision not to include these countries in the initial negotiations. They did this partly because of concerns that these countries would favor a one-country/one-vote arrangement similar to the International Telecommunication Union (ITU), but also because they knew that the main opposition to a global system would come from key countries in Western Europe, which were interested in protecting colonial cable and radio networks as well as potentially building their own regional satellite communications systems. The United States took for granted the support of countries in the Global South for a global system that would be open to their involvement and that would allow them to break free from colonial networks.

Thus, to arrive at the interim agreements, which were valid until definitive agreements were authorized in 1973, US officials worked to convince countries in Western Europe to join the single system and accept the US vision. These efforts would entail interrelated negotiations involving domestic and foreign policy decision making, as well as technical and economic calculations. Through analyzing the negotiations leading to the 1964 interim arrangements, in this chapter, I specifically explore the contested origins of governance for satellite communications, focusing especially on how and why the

United States worked to globalize its domestic vision for satellite communications during the early 1960s through Intelsat. "Global" was the ideal, but it was always in tension with regional and national considerations.

I examine the negotiations during three distinct periods leading to the interim agreements. The State Department decided to prioritize key countries and not involve the United Nations (UN) during the first period, from September 1962 to April 1963, but also conducted informal, exploratory discussions with Western European countries and Canada. Simultaneously, the UK and France, along with other Western European countries, explored national and regional policy options. Led by technical and administrative experts with telecommunications organizations, Western European countries increasingly coalesced as a regional bloc during negotiations with the United States. During the second period, from May to December 1963, foreign offices agreed to establish a new satellite communications organization for Western Europe, while prioritizing a common concern about contributing components manufactured by Western European companies. Comsat attempted to conduct bilateral negotiations with individual European countries but faced resistance from the US State Department and the new European organization, which both favored multilateral arrangements. US and European participants conducted formal negotiations leading to the 1964 interim agreements during the third period, beginning in February 1964. To bolster its position, the United States also included Japan, Canada, and Australia in these final negotiations. The support from this regional Pacific bloc played an important role in helping the United States counter an effective European bloc. The Kennedy administration's vision for a global system thus was implemented through not only domestic legislation but also negotiations with other countries involving global, regional, and national considerations.

Initial Planning during the Fall of 1962

The 1962 Communications Satellite Act only provided general guidelines for organizing the global system. The legislation was open ended or ambiguous about many specific considerations. William Carter, special assistant for international space communications in the State Department, who played a key role in the early international negotiations during fall 1962, later recalled that "there was certainly no agreement, and very little thought as to what the international arrangements would in fact look like at the time the Act was passed." According to Carter, "It really was a world to be made within the framework of very broad, and very encouraging, open, sharing, kind of policy words that were in the Act." Government officials had given "very little thought as to what the international arrangements would in fact look like. [T]here really was no plan at all. It was all to be made."[2] Although the act called for foreign participation, it did not detail the form of this participation. The United States would need to negotiate with other countries on specific issues, but beforehand officials with the US government and Comsat would need to decide their own positions on whether other nations should jointly own the system or participate in such issues as management, manufacturing components, and system design. Traditionally, private US companies, especially AT&T, had negotiated agreements for international radio and cable with telecommunications administrations of foreign governments, generally without the active involvement of foreign offices. Since the satellite legislation established Comsat as the chosen instrument of US foreign policy, the new organization would have to consider not only traditional business and technical planning, but also political calculations.

Before the initial agreements with foreign governments could be finalized, officials would need to clarify the relationship between Comsat and other agencies of the federal government—particularly the State Department, the Federal Communications Commission (FCC), and NASA. The satellite act stipulated that NASA should provide Comsat with technical assistance, but it did not clarify whether the government should subsidize a private company with monopoly control. The legislation was also unclear about the specific responsibilities of the State Department in the foreign negotiations. Would Comsat mainly deal with telecommunications administrations or with foreign offices in separate agreements? These tensions and ambiguities had to be worked out before the establishment of interim agreements allowing most of the actual construction to begin.

Although specific details involving the international system needed to be clarified through negotiations, the Kennedy administration's ideal of a truly global communications satellite system open especially to the involvement of third world countries was clear. Kennedy's ideal vision not only emphasized the importance of thinking in global terms but also the need to avoid privileging specific communications space, especially when it was defined commercially in terms of profitable communications traffic.

Despite the UN having passed a resolution declaring general support for global satellite communications in 1961, US government officials did not formally consult with other countries about implementing this resolution before passing the Communications Satellite Act.[3] The State Department first consulted with the UK and Canada in September 1962. A special Commonwealth Conference on Satellite Communications had instructed the two countries to discover US intentions.[4]

When Britain and Canada first asked the United States to discuss global satellite communications in fall 1962, the US State Department did not know how to respond.[5] Staff member William Carter claimed there were "no positions, there was not even conceptual papers as to what the international organization might look like or how we should get there."[6] But during an initial meeting in September, State Department and other US government officials agreed on two key issues.

First, they decided that the UN, especially the member agency the ITU, was not a "satisfactory" forum for planning or managing the global system. The ITU operated according to the principle of one nation, one vote, which seemed inappropriate for the new international satellite communications organization and the private company established by the 1962 Satellite Act, Comsat.[7] Nonaligned countries from Africa, Latin America, and Asia had been increasingly asserting their independence at the UN during this period, and although their participation in the global system would be important later, the United States wanted to first develop a general framework that would guide their involvement. The United States was thus reacting against the increasing assertiveness of Global South countries at the United Nations. The United States feared that if the UN was involved in managing or operating the global system, these countries could use the UN as a "forum to attack the system as 'a rich man's club'" or act as a bloc to demand special treatment, which might prevent rapid initial development of the system, a priority for the Kennedy administration.[8] Finally, the Americans worried that if they conducted negotiations in the UN, the Soviets might use the format to delay the global system. Unlike most Western countries, which were major users of international communications, the Soviets mainly needed long-distance communications for internal needs. Arnold Frutkin, director of international programs at NASA, pointed out, "The intrinsic needs of the Soviet Union are not the same as the US, UK, and Commonwealth countries in general. Essentially, they have internal needs. They are lacking any real motivation. A system could operate without the USSR or its consent."[9] Since the Soviets would not have a practical reason to support the development of global satellite communications at the UN, they likely would oppose it as a US initiative.

Second, US government officials decided that initially they needed to negotiate with a "core group of countries" especially interested in international communications "in order to make progress in defining the international arrangements." This decision implied that only heavy users of international communications would play a major role in initial negotiations.[10] Although the 1962 Communications Satellite Act called for "broad international cooperation," directing "care and attention . . . to economically less developed countries and areas," it did not clearly mandate the immediate or near-immediate internationalization of the system.[11] During an initial meeting with the UK and Canada in September 1962, R. N. Gardner, deputy assistant secretary of state for international organization affairs, argued that although the United States used "the name of 'universality' so as to get the cooperation of certain nations," it would "be principally concerned" with the nations of the "North Atlantic Community." Seeming to downplay the role of developing countries, he told British officials that the phrase "global and nondiscriminatory" in the 1962 Satellite Act "is a symptom of the lesser developed countries' desire to participate."[12]

British Foreign Office official Ronald C. Hope-Jones also supported a view about the role of developing countries that was becoming dominant within the US government: Although "underdeveloped countries" should be "allowed some voice" later, their participation "should be deferred until the world organization is reasonably sure of where it is going conceptually."[13] The British likely felt they could represent the interests of many of these countries because they were former colonies now part of the Commonwealth. And the United States probably wanted to avoid including former European colonies in the initial negotiations because France and Britain might view this as interfering with their interests.[14] Not deferring to France and Britain when dealing with their former colonies also might have unnecessarily emphasized British and French weakness in international relations following decolonization.

The UK and Western Europe Organize for Space Science and Technology

Britain and France were the two main countries in Western Europe interested in developing independent high-technology industries connected to space exploration and research. By the early 1960s, Western European countries had already begun to make plans to develop space capabilities. They recognized that they needed US technical assistance, but they also wanted to develop the independent industries necessary for space exploration and military rocketry.

Many Europeans viewed the technological developments connected to space research as an important new source of economic competitiveness. During the late 1950s and early 1960s, economists increasingly stressed the significance of science and technology for economic development. Countries in Western Europe, in particular, expressed fears about the economic implications of a growing technology gap with the United States based on electronics, nuclear energy, rocketry, and other modern industries.[15]

Important industry and government officials in Western Europe wanted to actively develop independent space-based industries, rather than wait "expectantly and pathetically for space crumbs which may fall from the rich man's table."[16] The United States opposed the transfer of technological innovations related to the space program not simply for economic or political reasons but especially because of the potential military applications. Perhaps most important, the government wanted to prevent European countries, especially France, from developing an independent nuclear missile force.[17]

Britain and France came together with other countries to establish the European Space Research Organization (ESRO) and the European Space Vehicle Launcher Development Organization (ELDO). In 1962, Britain, France, West Germany, Belgium, Italy, the Netherlands, Denmark, Sweden, Switzerland, and Spain signed an agreement establishing ESRO, a civilian agency committed to the peaceful pursuit of scientific investigation. During that same year, the first six of these countries along with Australia signed the ELDO pact. Forty-seven companies and trade associations—representing aerospace, electronics, and other industries—established a third European organization, EUROSPACE, in 1961, with a goal of creating a European industrial organization that would pool resources and provide expert advice on space programs. The creation of these institutions (especially ELDO and ESRO) reflected the efforts to achieve European solidarity following the Treaty of Rome and the establishment of the Common Market. ELDO was also organized to continue development of and conversion to civilian use a military missile, Blue Streak, canceled by the British government in April 1960.[18]

British officials had begun to seriously consider becoming involved in satellite communications as early as 1960.[19] The United Kingdom had long held a leading position in international communications, using both cables and radio to connect Commonwealth countries. A global satellite communications system threatened British control of Commonwealth communication. It would, for example, allow nearby locations well removed from London—for example, certain locations within India—to communicate directly instead

of having to route conversations through London as the hub of the Commonwealth network. The British Post Office Department, the main government
agency responsible for civil telecommunications in Britain, initially believed
that the Commonwealth should consider constructing its own system.[20] It recognized that satellites might threaten the Commonwealth cable system, but
officials believed the new technology would likely complement rather than
replace overseas cables, just as shortwave radio had complemented international telegraph cable systems in the 1920s. During fall 1960, the Post Office
accepted an offer from NASA to participate in satellite communications experiments. The government finished construction of its own ground station
during summer 1962.[21]

Some officials hoped this work would support a new industry involving
the sale of ground stations to other countries for use with a commercial system. When Hope-Jones heard that Pakistan was establishing an experimental
space communications ground station, he told an official with the Commonwealth Relations Office that he assumed "the equipment is coming from the
States, and that the engineers who are being sent abroad to receive training
are also being sent to the States." "We have always had it in mind," he wrote,
"that the establishment of a satellite communications system would provide
us with a good opportunity for exporting the necessary ground equipment
of one sort or another." He worried about the country "missing the bus" and
believed that they needed to act at once to "exploit this potential market."[22]

The UK Post Office had responded to President Kennedy's call for a global
satellite communications system by holding a special conference in London
in spring 1961 for countries belonging to the Commonwealth telecommunications network. A Post Office representative argued that "unless the UK
worked quickly, there was a danger" that the country would lose its "profitable place as a world telecommunications carrier."[23]

To provide leadership for the Commonwealth and to formulate official government policy, British postal officials organized an interdepartmental working group. As in the United States, different government agencies in Great
Britain disagreed about whether they should treat satellite communications
as a new form of international communications, demanding new political
and institutional arrangements, or simply continue to use traditional arrangements previously developed for international cable and radio. During meetings of the working group early in 1962, the Post Office and the Ministry of
Aviation both argued for developing an independent satellite communications system involving Commonwealth and Western European countries.[24]

Both agencies emphasized the political implications of the different technical systems being considered by the United States. Experts in these departments believed only a geosynchronous system would provide sufficient coverage for the needs of all Commonwealth countries, located across the entire globe. They worried that since the United States wanted to develop a system as quickly as possible, it would choose the least complicated one—a system of nonsynchronous medium-altitude satellites designed mainly to serve the needs of North America and Western Europe.[25]

British government officials realized that because of the US lead in space research, the Americans would inevitably develop a system first. But based on studies it had conducted, the Post Office Department believed a second system would be needed by the 1970s. It initially favored only minimal involvement in the US system; this would allow British and European industry to develop the capability to launch a profitable independent system. Officials in the Ministry of Aviation were mainly interested in finding a role for the development of the European launcher (the ELDO program). They argued that this program was needed because although the United States would be willing to launch European military and scientific satellites, it would be less willing to agree to launch commercial satellites developed by Europeans.[26]

Largely because of the efforts of staff member Hope-Jones, the Foreign Office eventually convinced the Post Office and most members of the working group to support US efforts to develop a single system.[27] Hope-Jones convinced other officials in the government that "American thinking" was "not nearly as far advanced as is sometimes supposed" and that the United States might still consider a system that would more favorably accommodate the needs of the Commonwealth.[28] The Foreign Office not only wanted to support a major ally but also believed that working within the single system would secure maximum advantage for British industrial and communications interests. "The quickest way of getting technical experience," Hope-Jones argued, would "be to co-operate operationally in the establishment of the original system."[29] The Commonwealth conference agreed with Hope-Jones's evaluation that a "purely Commonwealth system is out of the question, as it could not pay its way." Representatives from the Foreign Office also successfully convinced other members to pursue greater cooperation with Western European countries to maximize the Commonwealth's "bargaining position" with the United States. If the Commonwealth and Europe "can be persuaded to stick together over this," Hope-Jones argued, "they will be in a strong bargaining position, and it might prove possible to reach agreement that we would all co-operate fully" with the Americans "in the establishment and ownership of the first system, on the understanding that we would take the lead in establishing the second system when the need for this arises."[30]

The London conference instructed Britain and Canada to conduct exploratory talks with the United States and countries in Western Europe. The State Department met with representatives from these two countries in October 1962 and reassured them of the country's intention to accommodate the needs of the Commonwealth. The department also sent a team of officials to several European countries to brief appropriate representatives of US progress and plans.[31]

In general, two technical institutions representing different models for organizing space interests inspired the European space program's founders: the high-energy physics laboratory built during the early 1950s near Geneva known as CERN and the European Atomic Energy Community (Euratom), established in 1958 to create a common European market for nuclear energy. Although both organizations reflected the broad desire to integrate European institutions, CERN maintained a political separation from the supranational European institutions created during the 1950s, the European Coal and Steel Community (ECSC) and the European Economic Community (EEC). Euratom, by contrast, was created alongside these two institutions, with which it shared a Parliament and Court of Justice.[32]

The founders of ESRO were mainly committed to the CERN model, but because of the national security implications of launchers, national governments played a more important role establishing ELDO. ESRO's founders worried that if Western Europe did not pool its resources to support a space program, its citizens would remain "mere spectators of the grand endeavors to the East and West of our continent," but they wanted scientists in charge, free from political, military, or industrial meddling. Although ELDO was not connected to the EEC, Britain's involvement in ELDO was linked to its unsuccessful effort in 1961 to join the Common Market. Britain's ELDO involvement supported its bid to join the EEC by demonstrating "European credentials" to continental countries, especially the French.[33]

Great Britain's decision to cooperate with EEC countries also needs to be understood in the context of decolonization, which had weakened the country internationally. Increasingly, the British sought to pool limited Commonwealth resources with the economically powerful emerging alliances in Western

Europe. Western European governments, however, tended to oppose having Commonwealth countries join European collaborative efforts, especially as members with equivalent standing.[34]

France and the Decision to Use CEPT

France also worked to develop an official state policy for satellite communications (national and global). French technical experts had worked closely with AT&T engineers in early experiments with the Telstar satellite. AT&T relayed the first live transatlantic television broadcasts between ground stations in Andover, Maine, and Pleumeur-Bodou, France. The French director of the National Telecommunications Research Center (CNET), Pierre Marzin, believed these experiments had helped train more than one hundred French experts "who could participate in developments on a knowledgeable basis." Although French officials with the Foreign Office agreed that Europe should cooperate with the United States, they insisted, "Europe should not play the role of 'little boy' but seek to ensure the possibility of building satellites and perhaps launching these when she was able to do so."[35] Some French officials believed the country should avoid becoming dependent on a global system likely dominated by the United States. A similar motivation led the country to simultaneously develop a separate nuclear arsenal.[36] French government officials were particularly interested in building up French industry by supplying components to the planned global system. British diplomats reported that France would likely treat "the question as an entirely political matter, linking it to General de Gaulle's concept of 'a European Third Force.'"[37]

The French worried especially about the cultural and political implications of television transmissions from satellites, particularly the potential for broadcasting from satellites direct to individual homes, which could allow foreign countries, especially the United States, to inject ideas, propaganda, and English-language programming into former French colonies and undermine French influence. In early December, the British embassy in Paris reported that a key member of the French Foreign Ministry, Jean de la Grandville, was "obsessed by the television angle." He feared that "in the quite near future" US television programming in Africa would "give the Americans an enormous political advantage."[38] De la Grandville believed that the dramatic success in summer 1962 of the first live television transmissions across the Atlantic using Telstar "had completely changed the dimensions of the problem."[39] But British officials criticized de la Grandville for apparently not realizing that direct broadcasting from satellites to individual sets "cannot be done now" and would not likely become a routine service until the 1990s at the earliest.[40]

France's discussions with Carter and other US officials in late 1962 did not allay concerns that the United States would not "allow other countries a worthwhile form of participation."[41] France (and independently Sweden) requested that the Telecommunications Commission of the European Conference of Postal and Telecommunications Administrations (known by the French acronym CEPT) consider a joint effort to finance and organize satellite communications during a special meeting in December 1962 in Cologne. The French wanted Western Europe to organize a regional satellite communications organization that would serve a function similar to Comsat in the United States. Nineteen countries, primarily in Western Europe, had established the CEPT in 1959 to coordinate state postal and telecommunications organizations (PTTs). The Council of Europe had first proposed a common European postal organization as early as 1951, but the support of the six founding members of the EEC in 1958 was a crucial factor in its establishment.[42]

Representatives at the Cologne meeting of the CEPT authorized the establishment of an ad hoc committee primarily to study the technical aspects of the system in preparation for discussions with the United States. The seven-member committee was composed of officials from six Western European countries and a separate official representing the three Scandinavian nations. The CEPT delegates at the Cologne meeting did not indicate that their respective countries favored a separate system. But the CEPT was a traditional institution representing the technical and business interests of telecommunications administrations. In most cases, individual governments had not yet thoroughly considered the political implications of the US proposal. The British Foreign Office, in particular, did not think it was appropriate to continue to rely exclusively on this traditional institution.[43]

Most members of the British government opposed the French proposal for a new European organization supporting satellite communication.[44] Hope-Jones believed that the UK could consider supporting a loosely organized group, but he did not "welcome the creation of a closely-knit European organization of a kind which would speak with a single voice."[45] His position reflected increased tensions between the United Kingdom and France during this period. President Charles de Gaulle vetoed Britain's bid to join the EEC in January 1963, immediately after the Cologne meeting, and had been sending mixed signals about the application. As a representative of the Foreign Office, Hope-Jones supported the Commonwealth communications network and valued the special relationship with the United States. De Gaulle opposed Britain's membership in the Common Market partly because he first wanted the British to break free from the United States and the Commonwealth.[46]

The British view generally prevailed at the Cologne meeting; the representatives did not agree to establish a major new organization.[47] They favored European participation in the global system being developed by the United States, but partly because individual delegates mainly represented technical or administrative offices, they did not indicate whether their respective governments preferred a separate system.[48] French officials warned other countries, especially the British, that President de Gaulle would make the final decision about French participation in the proposed global system. The use of the regional organization, the CEPT, was particularly important because it underscored the fundamental tension between regional and global considerations in satellite communications.

At the beginning of 1963, the State Department stepped up efforts to prepare for serious negotiations by trying to obtain information from sympathetic foreign officials about European planning. In February, the director of the Office of Transport and Communications, Edward A. Bolster, asked the civil air attaché at the London embassy, John S. Meadows, if he could "conceivably find more friends within the UK Government for our point of view."[49] Whether Meadows responded by looking for someone is unclear, but soon after this request, the British official in the Foreign Office who had been advocating European participation in the US global system, Hope-Jones, began to provide embassy officials with detailed briefings about the "power struggle going on" in the British government.[50]

Hope-Jones stressed that the British government still had not made a final decision. Late in February the cabinet received several proposals "worked up by different agencies." The most important proposals came from the three main agencies interested in satellite communications. The Foreign Office's recommendation supported US plans. The Ministry of Aviation was mainly interested in receiving assurances that future planning would involve ELDO and the Blue Streak launcher. It favored participation in the single global system, but only if the other participants agreed to eventually use the European launcher to place communications satellites in orbit. The Post Office still believed that the government should consider developing a second system that might eventually be integrated into the primary system. Hope-Jones reported that nationalistic statements in Congress and in the US press, which seemed to indicate that the United States would seek a dominant role in the system, had weakened the Foreign Office's position. Specifically, members of Parliament and the cabinet feared that these statements indicated Comsat might "seek to continue its initial predominant role in the world system even after capabilities are developed by other countries." He recommended that the State Department make every effort to reassure members of Parliament and other government officials that the United Kingdom would have an "adequate opportunity over the long haul to participate substantially in the world organization." This would greatly strengthen the "Foreign Office hand." Hope-Jones also worried that "overly sensitive preoccupation with nationalistic objectives" by "underdeveloped countries" would likely "hinder the establishment and expansion of the satellite system."[51]

Sympathetic Canadian embassy officials in London also informed the United States about internal British political developments. The Canadians strongly supported US efforts, but because of their Commonwealth connections, they also felt a sense of loyalty to the United Kingdom. In March 1963, Canadian officials relayed to US diplomats their concern that "in the face of pressure mounting in favor of a separate system," some British government officials were losing confidence in the "single system concept." They worried because the British would likely pressure Canada to join a Commonwealth system; however, the Canadians reassured the United States that they had made clear to the United Kingdom the overriding advantages of a single system and that "a British decision now in favor of a separate system would confront us with difficult and conflicting considerations." The US embassy in London believed that "the Canadian initiative in reporting so fully to us on their actions confirms the previous indications we have received of their strong desire to cooperate with the United States."[52] British officials understood Canada's close ties to the US space program and how this resulted in conflicting loyalties. Hope-Jones acknowledged that the Canadians were "ahead of us in satellite engineering, having built their own satellite (Alouette)," and had been working closely with NASA, which had launched the satellite the previous year.[53]

The ad hoc committee established by the CEPT met in Paris during the middle of March to consider European participation in the US project and whether Europe should, at a later date, establish a complementary system. The British, French, and German foreign offices tried to convince the CEPT leadership to allow their representatives to participate in the deliberations of the ad hoc committee, but "at the last minute" the president of the CEPT decided that the charter did not authorize such involvement.[54] The president believed representatives could draw a sharp boundary between technical issues and political dimensions. The committee mainly tried to deal with narrow factual issues, such as whether the volume of traffic during the 1970s would be large enough to justify a second system.[55]

The committee agreed that countries in Western Europe should participate in the US system and that they should negotiate as a bloc instead of individually (at least initially through the CEPT), but the lack of knowledge about specific facts and the unrealistic instructions to ignore interrelationships between technical issues and foreign policy implications prevented the members from making definite recommendations. According to one participant, "a detailed examination showed the concept [of European participation in the global system] to be vague and discussions rapidly brought out complications." As a result, the committee decided to submit a list of specific questions to the United States about its plans for the system.[56]

Comsat Negotiations and Intergovernmental Meetings in Europe

National considerations represented by bilateral discussions were also important and were in tension with regional and global considerations represented by multilateral negotiations. This was especially clear in the conflict between the State Department and Comsat. During spring and early summer 1963, while Western European countries were becoming better organized, the newly created Communications Satellite Corporation (Comsat) started to consider the global system. Comsat initially indicated that it would pursue international negotiations using traditional models developed by AT&T and other companies—negotiating directly with individual foreign countries with minimal involvement of the State Department. But the State Department had difficulty getting both Philip Graham, the first chair of Comsat's board of incorporators, and Leo Welch, Comsat's first CEO, to agree not to usurp the authority of the department in conducting foreign negotiations.[57] State Department officials believed that although the Communications Satellite Act was not entirely clear about the proper relationship between the corporation and the department, the legislative history indicated that Congress had wanted to give authority to the secretary of state as the president's representative.[58]

To avoid conflicts with the State Department, during May 1963, when Welch and the president of Comsat, Joseph V. Charyk, planned the first foreign trip on behalf of the corporation to discuss the organization of the system with Canada and European countries, they told the State Department that their team would "limit the discussion to technical matters." Initially, they convinced the secretary that "it was both desirable and possible to make a strict separation between technical and organizational or policy matters." Although the secretary expressed reservations about "whether he would adequately be fulfilling his and the President's statutory responsibilities if there was no government participation in these early stages," he grudgingly agreed that under these conditions, State Department personnel did not have to be present during the discussions. But he also received assurances that Welch would inform the foreign operating agencies that he was not "speaking for the United States Government." Welch agreed that if Foreign Office representatives were present at any of these meetings, he would invite State Department personnel to participate.[59] When the two executives held talks with Canadian government officials from the Department of Transport at the end of the month, they informed the US embassy in Ottawa that they had only dealt with such "general technical problems" as "the estimated timing of an operational system, the comparative financial implications of the synchronous and medium altitude systems, and the problems of maintaining technical development along compatible lines."[60]

But several US government officials questioned whether it was realistic to assume that Comsat personnel could make a sharp distinction between technical and nontechnical considerations. The issues discussed with Canada seemed to involve a combination of technical, economic, and other issues. Even if the State Department accepted that Comsat could discuss with foreign telecommunications administrations "technical" issues broadly defined to include narrow economic or business considerations, at least one official still argued that it was "extremely difficult to make satisfactory distinctions at the present time between 'business' and 'political' subjects."[61] This official as well as a number of others objected to the arrangements the secretary of state had made with Welch and Charyk. They pressured the State Department and other government representatives to clarify the "proper relationship between the Department and the Corporation" as they "move forward nationally to implement the global system called for by the Communications Satellite Act of 1962." Officials found especially troubling critical comments leveled at the Comsat executives by a member of the FCC on May 24. The FCC official claimed that Welch and Charyk, despite their previous statement to the secretary of state promising that they would place sharp limitations on their foreign discussions, had told him that they "did not feel themselves under any real restrictions to discuss only technical matters."[62]

Indeed, during "bilateral discussions" in Europe with telecommunications officials from the United Kingdom, France, the Federal Republic of Germany, and Italy between May 27 and June 1, the two Comsat officials discussed all aspects of planning for the global system. They not only dealt with material, or equipment, issues such as the problem of synchronous orbit injection, the development of spin-stabilization techniques, power sources for satellites, the reliability of available launchers, and the expected lifetime of equipment, but also discussed plans for four levels, or categories, of participation for different countries. In the first category, the corporation proposed including major countries interested and able to serve as co-owners by contributing capital for construction and operation, in proportion to their use of the system. Countries in the second category might be unable or unwilling to compensate for expenses individually but interested in contributing capital as part of a group of countries using a common ground station. Nations in the third category would use the system by leasing channels but would not contribute capital. The Comsat executives indicated that poorer countries in the fourth category interested in using the system but unable to contribute might be able to borrow ground stations, secure loans from the World Bank, or receive direct aid from wealthy countries such as the United States or France. The two officials specifically indicated that these countries should not expect assistance from the company.[63] They stressed that Comsat would make decisions based on "purely commercial" considerations.[64]

Welch and Charyk also informed the European officials about their preliminary plans for the construction of a global system. They hoped that by July 1963, the United States would have "some basic data on the minimum characteristics needed for the system." They planned to input these data into computers to compare the economic advantages of different systems and finish the preliminary analysis by November 1963. By May 1964, they expected to make a final decision about the type of system to adopt, the future timetable for construction, and the level of participation and capital contributions of different countries as well as those of individual stockholders. Operation of the first elements of the system would begin in 1967. The Comsat officials informed the Europeans that they expected a central authority located in a single command and control center to manage the system, but that the European participants would take an active role in decision making through membership on a board of directors.[65]

The French response to Comsat's plans emphasized the country's extensive experience with international communications. French officials pointed out that when evaluating the problem of time delay in voice communications using geosynchronous satellites, experimenters needed to take account of the "rapidity of the language being used." They had found in years of experience with telephone communications that Africans who speak extremely fast have more difficulty with even slight time delays. Further, the French expressed that they did not believe that television transmissions across the Atlantic would be an important source of revenue for the satellite system, mainly because of the significant time difference. Pierre Marzin, the director of the French National Telecommunications Research Center, did not believe "real-time television" would be extremely popular or cost effective, at least in the short term, compared to taped television programs transported across the ocean.[66]

In June, Marzin also warned Comsat officials not to overemphasize the single global system; he reminded them that it "had become a political hotspot in some quarters." The French representative cautioned that for some Europeans, a single global system still "implied complete American dominance." He reported that French government officials in charge of space research had told President de Gaulle that they still planned to eventually establish a separate system. Marzin admitted that although he would be satisfied if his country launched "one good operating satellite" in ten years, "politically he must tell the French Government that the Europeans can launch a satellite by 1967." He did not believe that other French officials realized the complications and expenses of establishing a satellite communications system.[67]

Instead of emphasizing that other countries, including developing countries, should be involved in the initial stages of planning, US officials argued that all countries should be "able to participate in the system as soon as practicable."[68] Although the State Department supported this decision, it generally opposed Comsat's efforts to ignore political considerations. Carter later recalled a "state of dynamic tension" between Comsat and the State Department during this period.[69]

As a commercial company, Comsat tended to reflect the traditional view of "telecommunicators" such as AT&T. The company sought to make decisions based on commercial considerations and tried to limit the role of the State Department and foreign offices from other countries. According to Carter, "the telecommunicators didn't like foreign offices and foreign ministries mucking around with" their established practices.[70] When Comsat executives conducted their own negotiations with Europeans, they emphasized that the global system should be based on "good business rather than primarily political considerations."[71] The State Department accepted the company's emphasis on traffic volume and agreed to first seek an interim agreement with a "key or nucleus group" of countries that included not only Western European nations but also Canada, Japan, and Australia. According to a US study, more than 90 percent of intercontinental commercial communications traffic "either originates or terminates within the United States or members of this group."[72] Under pressure especially from Comsat, the US government also agreed that financial assistance to developing countries to aid their participation "would be the responsibility of national and international sources of capital assistance and not of the entities participating in the satellite system."[73]

Partly in response to some US government officials' concerns about the State Department's lack of involvement in Comsat's bilateral discussions in Europe, on June 5, 1963, government officials decided to establish an Ad Hoc Communications Satellite Group, directed by Nicholas deB. Katzenbach, attorney general, and Jerome Wiesner, special assistant to the president for science and technology, to coordinate the work of government agencies in dealing with Comsat.[74] Members of the ad hoc group pressured Comsat executives to avoid making definite and inflexible arrangements regarding such issues as management and ownership before consulting thoroughly with all Europeans. US officials opposed Comsat's policy of wanting to include only countries that were "able and desiring to make a substantial contribution" to the system to expedite its establishment. An initial Comsat memorandum detailing operating principles submitted to the ad hoc committee did not seem to uphold the "principle of non-discriminatory access" emphasized in the Communications Satellite Act. Government officials convinced Comsat to make "an effort to enlarge the means whereby countries not included in the original Space Group could subsequently acquire ownership in the Space Segment." But even after the corporation drafted a new memorandum in summer 1963, government officials were still not convinced countries would be able to acquire ownership in the system after the operational date. US officials were especially worried that Comsat executives seemed to ignore "extensive reporting of European attitudes and actions," which clearly indicated they did not want to be treated as junior partners and preferred multilateral agreements.[75]

US and European representatives continued informal discussions during 1963. But Western European countries did not begin formal negotiations with the United States until February 1964. Before these formal negotiations could begin, the interested European countries needed to conduct their own discussions to decide on a formal political response.

First Intergovernmental Efforts in Europe to Address Satellite Communications

The first official intergovernmental political effort in Western Europe to consider satellite communications, involving foreign ministry rather than telecommunications or space-oriented officials, took place in Paris in May 1963. The Europeans then held two additional major intergovernmental meetings in 1963—in London in July and Rome in November—as well as a series of smaller meetings of special committees established to investigate technical, financial, and administrative issues.

During the three intergovernmental meetings in 1963, officials debated whether Western Europe should pursue a separate satellite communications system or exclusively join the US global effort. They also discussed a related proposal to treat the separate system idea mainly as a negotiating tactic. An Italian Foreign Ministry source informed US diplomats in May that the French and British were "obviously most interested in pushing ahead as rapidly as possible on development of European projects aimed at improving their technological position and strengthening their hand for later negotiations with us."[76] During the May meeting in Paris, the French and the British agreed that Europeans should use the threat of an independent "subsystem" to obtain the maximum number of concessions from the Americans. At least initially during the meeting, the French favored a separate system not only as a negotiating strategy but also as official policy.[77] By the July meeting in London, however, the French position had changed. Although the country still had "strong reserves" about the American proposal for a single global system, they recognized their isolation on the issue and no longer "openly" opposed it.[78] By the November meeting in Rome, the Europeans had effectively abandoned the idea of using a separate system as a negotiating tactic. The chairman of one special committee who also served as an official with the Dutch Foreign Ministry, S. Meijer, acknowledged in October that the "separate system concept [was] dead as of now."[79]

A second major issue debated at the intergovernmental meetings was the type of organization the United Kingdom, France, and other Western European countries would adopt. Britain opposed immediately establishing a formal European satellite organization, favoring instead "some loose form of European organization" that would not require "elaborate negotiations" and would somehow provide for the country's "special Commonwealth connexions and responsibilities in the telecommunications field."[80] During the July meeting in London, however, the smaller countries—including Switzerland, Belgium, the Netherlands, and the Scandinavian nations—teamed up with France to create a formal regional organization to provide a "united European front" in negotiations with the Americans.[81] The organization was named the European Conference on Satellite Communications (CETS). The British reported that the West Germans "played little part" in the London conference. Ireland was the only other country besides Italy supporting Britain's call to hold off organizing a regional group with centralized decision making.[82]

Following the London meeting, Britain moved closer toward the general European position favoring a regional association.[83] Major divisions remained over whether the new satellite communications organization should become the main European representative of the planned international organization or should allow countries participating in the international organization to "apply for individual membership, speaking, however, with a single voice." Regional and local considerations remained in tension with the global ideal.[84]

A third major issue discussed at the three intergovernmental meetings in 1963 was Western European industry's desire for a US guarantee allowing different countries to support local manufacturing by supplying components to the planned system.[85] Hope-Jones told the United States in early June that Europeans "were thinking of some sort of 'infant industry' approach, during which some guaranteed share of procurement and research would be allocated to the European countries."[86] But Welch and Charyk consistently held to their policy of not taking into account commercial or business considerations. They refused to promise special treatment for European industry. Comsat would use equipment produced by European manufacturers only if it met required specifications at a lower price than other companies.[87] Although the Comsat executives refused to agree to allow Europeans to supply components on a "guaranteed basis," Welch did make a small concession during the May meeting in Paris, when he "left the thought" with the Italians that the international board of directors would make the final decision. The Europeans also won a small victory when Italian officials convinced Welch and Charyk to agree to allow European technicians to work with American firms three months earlier than the original Comsat proposal.[88]

Although by the July conference the European satellite communications group seemed to acknowledge that US companies would likely construct nearly all the components for the first generation of the communications system, all the countries—especially the smaller ones—continued to argue that Comsat should allow them to eventually participate as both users and manufacturers of later generations of the system.[89] Europe's Common Market institutions also sought to coordinate European industry, but they viewed economic integration as especially important for maintaining peace in Europe and achieving political integration. The satellite group, by contrast, was especially concerned about pooling resources and integrating economically to compete with US industry.

The fourteen earlier members of CETS and three new members—Cyprus, Greece, and Ireland—decided during a meeting in Rome in late November to form a new regional organization with a specific mandate to serve as a European partner working with Comsat in the establishment of an international satellite communications organization.[90] Delegates from Australia also attended as observers because of the country's membership in ELDO.[91] The conference directed the steering committee to assume responsibility for negotiating with US government officials acting on behalf of Comsat in future meetings about setting up the international organization. They invited the United States and Canada to attend an exploratory meeting in Rome early in February in preparation for the first official negotiating conference in March, which would work toward forming a provisional intergovernmental agreement. To emphasize the subcommittee's important new role, the conference renamed it "the Committee of Deputies."[92]

Although during the conference in Rome, the delegates discussed many of the practical issues that the subcommittees had explored with Comsat and the State Department during the previous few months, they were particularly concerned about the recent announcement by Comsat that it planned to begin experimental operations in 1965. The State Department suspected that the Europeans were interested in "attempting to delay implementation of an operational system" to "give them time to catch up" to the United States, giving a boost to European industry seeking to compete with US companies manufacturing components for the satellite system.[93] But a more significant motivation for Great Britain's desire to delay the system was the government's interest in modernizing and expanding the country's extensive submarine cable system. The British Post Office hoped to complete the installation of a high volume, transistorized submarine telephone cable by 1966, in time to meet anticipated transatlantic traffic needs. Although US State Department representatives suspected that "financial considerations influence[d] the British view to a large extent," the British argument that they had to have a "completely sure addition" to their telecommunications system did have some validity.[94]

Unlike the new cable systems, which could more easily be tested before installation, only preliminary experiments had occurred to justify claims that communications satellites could provide high-quality communications circuits at a low price. Telecommunications officials still worried, for example, about the potential problem of time delay and echo that might occur with synchronous satellites. If experiments with communications satellites failed to provide satisfactory service after officials had decided to delay production of the new cable, they would need at least two years to reverse the decision and complete installation of a cable system to meet the expected increase in customers' requirements.[95]

US officials who were worried about competition from cables recognized AT&T's important role. Approximately three weeks before the Rome conference, Ralph Clark, from the Office of Telecommunications Management, warned White House officials committed to the speedy development of global satellite communications that "the rapid expansion of the number and capacity of international telephone cables can pose a threat to the early financial success of any international communications satellite system." Clark specifically pointed to rumors about a planned project to install a new cable between the United Kingdom and Canada, almost doubling the North Atlantic capacity, as well as plans already well under way to produce a transistorized cable that would provide more than seven hundred channels for communications traffic between the United States and France. Clark emphasized that these foreign plans all depended on the support of the dominant international carrier in the United States, AT&T. He recommended that the Executive Office "take the lead" with the Department of State to discuss with the top-level management at AT&T the company's plans for expanding international communications capacity to make sure a commitment to new cable projects did not "sabotage the early success of the Communications Satellite Corporation." These events involving AT&T seemed to confirm some of the worries that liberal members of Congress had expressed during the debates leading to the passage of the Communications Satellite Act: if the government did not play a major role in the development of satellite communications, telecommunications organizations with an economic interest in overseas cables would be less likely to support the development of a global satellite system. To promote the speedy development of the global system, Clark argued, the government should not allow "unlimited competition" between the two technologies, which were both undergoing rapid development.[96]

Especially after the Department of State warned in early December that the Europeans were "far from satisfied that satellites should be chosen over cables," government officials pressured AT&T to take a definite position in support of communications satellites and make this position clearly known to Europeans.[97] AT&T executives gave a positive response in a letter to Comsat chairman Welch on December 6 and during a meeting ten days later with the acting secretary of state, George Ball, and the White House science advisor, Jerome Wiesner. The AT&T representatives reassured the officials that they believed a place existed for both submarine cables and communications satellites. They disagreed with the view expressed by Europeans of an "either-or-choice" and predicted that a dramatic growth in the volume of overseas requirements would justify the use of both technologies. The company placed a high value on having a "diversity of routes and facility types" from "the standpoint of day-to-day service integrity and from the standpoint of national security." The executive vice president promised that if Comsat executives and government officials could provide assurance of the availability, by at least 1966, of a sufficient number of high-quality satellite circuits to meet the expected increased demand in the North Atlantic region, AT&T would "prefer, where high capacity cables could be attractive, to use satellite circuits instead of placing additional cables." He also promised that in other areas of the world where his company had made preliminary plans to provide service through cables, notably between the United States and countries in South America, AT&T would immediately initiate discussions with its overseas partners to convince them to first use the expected satellite circuits before laying new cables.[98]

Government officials also pressured AT&T executives to accept an offer to attend the next meeting planned for early January of the ad hoc committee of the CEPT, which had specifically been organized to discuss the relative advantages of cables and satellites. State Department officials believed the January meeting offered "the last opportunity to make our case to European communications officials, whose recommendation will largely determine the position taken by European Governments at the intergovernmental caucus scheduled for early February." They believed AT&T's role would be crucial: "AT&T has the opportunity to sell the Europeans on satellites once and for all at this meeting."[99] Although the executive vice president of AT&T, James Dingman, had indicated in his letter of December 6 that his company was mainly committed to business considerations, notably providing high-quality service at low cost, when government officials met with him, they also stressed that the company had an obligation to uphold a national policy based on the imperatives of global Cold War by helping "develop satellite communication as an affirmative part of the historical leadership role of the United States, and as providing a necessary communication instrument to both demonstrate and implement that leadership." Ball reminded AT&T of the role of satellite communications in the global competition to use technology to demonstrate the political and economic superiority of American institutions. The AT&T executives only grudging accepted this argument. Frederick Kappel, the chairman of the board, who also attended the meeting with high-level government officials, related that he "was not too happy about the degree of political involvement." He thought that "some delay had occurred because too many people other than communicators" had been involved in the development of the global system; however, he did acknowledge that political involvement was "inevitable and that close collaboration between industry and government was required."[100]

The "talking paper" the State Department staff wrote to guide the acting secretary of state during his meeting with AT&T explicitly listed the points the company should tell Europeans. First, they thought AT&T should indicate that it "fully supports the national policies of the United States as defined in the Communications Satellite Act of 1962 calling for the development of a satellite system to provide global coverage as soon as practicable." Second, the company should strongly reassure the Europeans that in the company's judgment, the global system was "achieving sufficient progress to make it reasonably certain that satellite channels of the requisite quality and quantity will be available in late 1966 or early 1967." Third, AT&T executives should tell European representatives that they believe considerations such as cost, diversity, and flexibility "make it important that a satellite capability be developed expeditiously." Finally, the talking paper hoped AT&T would demonstrate to Europeans its commitment to satellite communications by indicating that it planned "to make a substantial investment" in the satellite corporation.[101]

Whether AT&T actually did tell Europeans these specific issues remains unclear. Kappel told Ball and Wiesner that "he did not believe the 1966 target date was realistic." Nonetheless, he did think "every effort should be made to meet it."[102] And AT&T agreed to attend the meeting. Perhaps more important, company executives conducted bilateral talks prior to the meeting with the different countries in Europe, in which they emphasized their commitment.[103]

After AT&T announced in fall 1963 that it planned a new high-capacity trans-atlantic telephone cable, the US government also feared that AT&T and the PTTs in Western Europe would use transatlantic cables rather than the global satellite system for international telecommunications. By December 1963 AT&T had committed to the planned global system and agreed to pressure the European PTTs to use satellite circuits before laying new undersea cables.[104]

The United States, Europe, and the Interim Agreements, February–August 1964

Exploratory talks in Rome during early February 1964 gave US officials their first opportunity to formally discuss organizing the global system with the entire European Conference on Satellite Communications. The United States satisfied the Europeans by sending an official government delegation headed by Abram Chayes, a member of the Harvard Law School faculty, with expertise in international law, who supported President Kennedy's campaign during the 1950s and then served as legal advisor to the State Department beginning in 1961. The State Department explicitly stated that his decisions within the delegation would be "final and binding."[105]

Chayes had to follow guidelines developed in cooperation with the State Department, the FCC, and other federal agencies involved in communications policy. The US government's Ad Hoc Committee on Satellite Communications coordinated; Katzenbach, representing the committee, and Chayes also met with key members of Congress—including Senators Pastore, Magnuson, and Fulbright—to ensure they approved a position paper and a draft agreement formulated to guide the US delegation's work.[106]

The guidelines laid out in these documents emphasized that the delegation should first explore the establishment of interim agreements based at least initially on the experimental synchronous satellite scheduled to begin operating over the Atlantic in 1965 before discussing—"in rather generalized terms"—the characteristics of the final, or "definitive," agreements based on a "fully developed global system." But the guidelines emphasized that the United States should prolong the interim agreements as long as possible. They stressed that the country should take advantage of its "monopoly of the boosters and of the communications technology" to establish as strong a position as possible for the final agreements. The guidelines reminded the delegation that Europe's "primary interest" would be to prevent the United States from continuing "to monopolize the system indefinitely."[107]

The guidelines also directed the delegation to explore ways other nations could participate in all stages of the system. US officials preferred having countries invest directly in the system rather than lease channels, but Comsat emphasized that if necessary, it could finance the system "entirely" and simply lease out channels to other countries. Reflecting a continuing commitment to the global ideal, the guidelines directed the delegates to take account of the interests of "countries in underdeveloped areas" unable to substantially invest in the early phases of the system.[108]

The Europeans brought up the issue of Soviet participation in the satellite system for the first time in December. A true global system would obviously need to include the Soviet Union, but US officials were ambivalent. Like the space program in general, the effort to establish a global satellite communications system needed to finesse the tension between Cold War competition for technological superiority and the propaganda value of cooperation. British officials agreed with the Americans that they should not try to bring in the Soviets until they had made substantial progress in designing and organizing the basic system.[109]

US officials also spelled out their preferred management and organizational arrangements. Most recognized that they would likely need to agree to an Interim Communications Satellite Committee, which would set policy for the initial basic system. Only countries investing a minimum amount of capital would participate in the interim committee. A country's level of investment in the system would determine their influence in decision making. The guidelines instructed the delegates to lay the groundwork for creating "as large a role as possible" for Comsat in the system's "day-to-day management." Finally, the delegates needed to stress the now definite US position that the country would not guarantee procurement for Europeans, at least during the interim period. Getting the Europeans to agree to these key aspects of the interim agreements would likely result in their pressing "hard for assurances on the ultimate organizational structure." The position paper warned that "this may be the price that they demand for permitting the Corporation to manage the interim system."[110]

In preparation for negotiating with the Europeans, the United States decided to include other countries to avoid "a purely European-American negotiation and possibly a European-American stand-off." The United States already had Canada's support, but they included Australia and Japan as two further potential allies. US officials arranged a meeting with Japanese and

Australian officials in Tokyo in January 1964 and, after a week of intense negotiations, gained their general support for the US position.[111]

The negotiations leading to the two interim agreements signed in July and August 1964 occurred during meetings from March through July in Rome, Montreal, London, and Washington, D.C. An intergovernmental agreement established a single commercial system that would provide global service "as soon as possible" and be "open to all nations on a non-discriminatory basis." Each government signing the agreement would also designate a "communications entity" (governmental or private) to participate in the system. For the United States, this entity was Comsat. A "special" operational arrangement between the communication entities designated by each government spelled out operational policies, management and administrative procedures, and financial understandings.[112]

The arrangements directed that Earth-station ownership would differ from ownership of the space system. Participants would jointly own the space system, including the satellites as well as the associated tracking and command and control equipment. Individual nations would own their Earth stations and Earth-station transmission facilities. Each government agreed that its entity would contribute a percentage of the estimated total costs for the space segment, approximately 200 million dollars over the first five years. Each country or group of countries contributing at least 1.5 percent would be represented on an Interim Communications Satellite Committee with "overall responsibility" for developing and operating the system. Comsat, as the manager of the system, would implement committee policies.[113]

One of the first major issues addressed during the final negotiations was Europe's opposition to Comsat's efforts to conduct bilateral talks. Following the meeting in Rome in February, Chayes discovered that he had succeeded in alienating the Europeans by pushing too hard for this unpopular position. European representatives believed Comsat was trying "to divide the European Conference." Small European countries in particular felt that their views were "considered to be of small significance" to the Americans. The French Foreign Ministry responded to Comsat's efforts by refusing to receive members of the US negotiating team during talks in March. A Dutch official reported that other members of the conference received news about this action "with great satisfaction."[114]

Comsat officials eventually agreed not to conduct separate negotiations. They backed down after meetings in early June when they realized that European solidarity did not mean Western Europe would "act as a bloc" in the interim committee "in antagonism to other members." The Europeans formally agreed that individual members should feel "free to act individually in matters [of] special interest to country or countries concerned, and to vote accordingly even if other European members were to take [an] opposing position."[115]

One of the most controversial issues and one of the last to be resolved during the final negotiations was the initial division of investment shares and the corresponding voting formula. Voting percentage was equivalent to investment or ownership quota. In using a weighted system of voting that favored the United States as the largest contributor, Intelsat followed the governance structure of the World Bank and the International Monetary Fund (IMF). The United States effectively used its economic weight and technological dominance for political gain.[116]

Western European countries initially requested a 40 percent US ownership share but agreed to a higher amount following negotiations with the United States. The United States preferred an investment share closer to two-thirds, which was calculated based on the country's share of international telecommunications traffic, but finally settled for 61 percent during a late-night negotiating session in June with leaders of the European delegation in the hotel room of a Comsat executive. They agreed that all Western European countries together would have 30.5 percent (half of the US ownership), with 8.4 percent for the UK and 6.1 percent each for France and West Germany. Canada, Japan, and Australia would divide the remaining amount, 8.5 percent.[117]

The European representatives originally wanted to see future reductions in the US share as new participants joined the system but finally accepted the US proposal that as new countries joined the consortium and received an appropriate quota, all the original participants' percentages would be reduced on a pro rata basis. The relative relationship would thus remain nearly the same. Crucially, however, the parties agreed that Comsat's quota would never be less than 50.6 percent, assuring the United States a majority while the interim agreements held.[118]

The related issue of voting structure was finally decided during a July meeting in Washington, D.C. Although the United States had a clear voting majority and the resources to go it alone after Comsat raised 200 million dollars through the sale of stock, American officials recognized that politically they needed other countries to concur with important decisions about the construction of the global system. But they were also concerned about "European obstruction" hindering developments.[119] US officials initially proposed that at least two other countries needed to vote with them on major decisions. The Europeans rejected this suggestion because the two countries could have very small voting percentages, or the United States could simply rely on Canada, Australia, or Japan without the support of a single European member.[120] The Europeans initially proposed that the United States should need the support of at least 30 percent of the remaining votes. The head of the European delegation worried that the four "Pacific countries"—the United States, Canada, Australia, and Japan—could "act as a bloc" against Europe. Although they generally supported the US proposal, the Canadians and the Australians urged the United States to negotiate a new position with the Europeans.[121]

All parties agreed on a complicated compromise for voting. If the interim committee could not "act unanimously" on fourteen key categories of decisions, then the United States would need the support of 12.5 percent of the remaining votes.[122] But if the committee did not act on certain cases within sixty days, then this amount would be reduced to 8.5 percent (Canada, Australia, and Japan's combined voting percentage). The United States requested this last requirement to prevent European obstruction. US representatives expressed satisfaction with the final voting formula. One analyst at the State Department claimed that the compromise formula would actually "in time" allow Comsat to gain more voting control than the original US proposal.[123] Importantly, although the European Conference negotiated with the United States as a unified entity, it accepted the British request to allow members to vote separately on an individual national basis. The British had been especially concerned about reserving the "right to express national views" because it "had commonwealth interests to protect."[124] The United States accepted the compromise voting formula partly because it believed it could exploit differences among the Europeans.

Procurement was another major concern. The Europeans worried that they could not compete for procurement contracts because of US industry dominance and because of the US demand that contracts be rewarded competitively based on the best equipment at the best price. Although article X of the intergovernmental agreement stated that the consortium should try to distribute contracts to the nations that had signed the agreement, it also emphasized that this would be done "provided that such design, development and procurement are not contrary to the joint interests" of the parties to the agreement. The United States managed to finesse the issue to satisfy the concerns of the original countries invited to the negotiating table, but the dominant position of the United States in procurement was clear.[125]

This chapter underscores the contested process that led to the first international agreement for the Intelsat global satellite communications system. The 1962 Communications Satellite Act had not provided specific instructions for establishing the global system. The organizational details needed to be worked out among several different groups and individuals. The first necessary step involved negotiations among US government agencies and Comsat. The 1962 legislation gave Comsat a foreign policy role in negotiations with other countries, and this led to tensions with the State Department especially. Comsat and the State Department tried to divide responsibilities by distinguishing between technical issues and political considerations, but in practice, when Comsat met with officials from other countries, they discussed both types of issues. Comsat thus practiced a version of the boundary work identified in chapter 3 with the 1963 Space Radio Conference. After US government officials resolved some of the early interagency conflicts, they made two key decisions about the global system. They decided not to involve the UN or the ITU, and they decided to conduct initial negotiations mainly with countries in Western Europe, partly because they were the other major users of international communications but also because they would be most likely to oppose a global system.

But Western Europe was not unified, and similar to the United States, different government agencies in individual countries needed to first formulate national policies. The same tension between "telecommunicators" (or in the context of Europe, the PTTs) and foreign policy officials in the United States also existed in Western Europe. Unlike foreign offices, the PTTs in Western Europe also tended to view the technology not as a radical new form of communications but as an extension of existing technologies on the ground. Further, Britain and France, especially their foreign offices, were interested in ensuring that communications satellites would support existing relations with their colonies and former colonies around the world.

Countries in Western Europe shared a common interest in wanting to use innovations connected with space exploration as the basis for new industries. They wanted to work with the United States to gain access to these innovations, but they also wanted to be able to develop independent industries. They realized early in negotiations with the United States that trying to develop independent satellite communications systems was not realistic and then settled on trying to convince the Americans to give them as strong a role as possible in the Intelsat system. France and many other countries in Western Europe wanted to negotiate with the United States as a bloc. Although the United Kingdom did not initially support this position, it eventually did go along with the establishment of a regional European organization, the European Conference on Satellite Communications (CETS), which had a mandate to negotiate details of the first interim Intelsat agreement. But the United Kingdom did win support for its position that individual countries should be allowed to also vote separately on individual positions.

Because the organization of Intelsat led to the establishment of the regional organization CETS, the global project actually reinforced regionalism in Europe.[126] With Intelsat, we see how regional and local issues were in tension with a global ideal. Further, the involvement of crucial countries outside Europe was shaped by their identification with regions dominated by the United States. Significantly, Canada prioritized its regional North American identification over its allegiance to the British Commonwealth.[127] And in the final negotiations leading to the 1964 agreement, Japan and Australia played a crucial role supporting the US position because they shared the Pacific Ocean region. Despite having to make compromises with countries in Western Europe during the negotiations leading to the establishment of Intelsat, the United States clearly dominated the new institution and was in a strong position to organize a true global system that would, on the one hand, specifically include as many countries in the Global South as possible and, on the other hand, generally support President Kennedy's original vision about using the institution to assert US geopolitical leadership around the world.

Conclusion

This book explores the establishment of the first global satellite communications system, a major infrastructure project with important implications for understanding the dominant role of the United States in the post–World War II world. As with other fundamental infrastructure projects, the Intelsat system has been underappreciated because its crucial role in facilitating international communications has been taken for granted. By uncovering the complex work involved in establishing Intelsat, we gain a better appreciation of not only how infrastructure systems are not inevitable or uncontroversial but also how they involve both technical components and organizational, political, economic, and business considerations. Decisions made during the Kennedy administration were crucial in the social construction of the system, and this study examines the full range of actors involved in decision making, not only elites but also lower-level administrators and agency officials who helped to negotiate domestic and foreign policy considerations.

By placing the origins of the large-scale technological system in the context of efforts by the Kennedy administration to use spectacular triumphs in space to demonstrate global leadership, the study extends the analysis of other scholars who have emphasized the global nature of the Cold War. The Cold War was not only an East-West struggle for military superiority but also an effort to use spectacular and useful innovations in science and technology to win over hearts and minds around the world, especially in developing and nonaligned countries, by showing practically and symbolically which nation had the better political and economic system and which country was more devoted to world peace.

As nonaligned countries gained influence during this period at international organizations like the United Nations and the International Telecommunication Union, North-South global tensions increased markedly. With satellite communications, we see how the Cold War could be simultaneously waged in both outer space and in the Global South. The idea of global satellite communications was both spectacular and practical; it could be used to show countries in the Global South in particular how they could benefit from the space race. And when these nations were convinced that they could participate, they were more likely to support US efforts to gain backing for international agreements to set aside valuable and scarce radio frequencies for all uses of space, including by the military.

Fighting a global Cold War was also increasingly "total," most importantly because of entangling domestic and foreign policy considerations in the United States. This issue was clear in the Kennedy administration's decision not to allow AT&T to establish a separate satellite communications system but to lead the world in the establishment of a single global system potentially benefiting all nations. Traditional domestic antitrust concerns became tied to national security issues, and this explains the Kennedy administration's decision to support alternatives to AT&T's communications satellite design and to work to block AT&T from extending its telecommunications monopoly into outer space. Officials in the Kennedy administration and key members of Congress worried that AT&T's design would give preferential treatment to major industrialized countries rather than lead to the establishment of a true global system benefiting developing countries in Asia, Latin America, and Africa. They also worried that allowing AT&T or other private companies to dominate satellite communications could be used to support anti-American Soviet propaganda. And government officials were concerned that AT&T's competing interest in developing undersea cables would prevent the company from rapidly establishing global coverage for satellite communications, which needed to include third world countries.

The interim agreement establishing Intelsat in 1964 resulted from negotiations the United States conducted with other nations that reflected global, national, and regional considerations. But this effort also underscored an inherent contradiction. Although US officials had placed great emphasis on creating a single global system that would benefit the entire world, especially developing countries in the Global South, the initial negotiations leading to the interim agreement only included major industrialized countries, especially nations in Western Europe. Thus, the notion of global that underpinned the organization of Intelsat was problematic. The system was global in the sense that all countries could join and participate, and the system was global in terms of the infrastructure covering the entire world. The Intelsat system achieved global coverage in 1969, when geosynchronous satellites successfully served the Pacific, Atlantic, and Indian Ocean basins. By the end of the 1960s, more than sixty countries belonged to Intelsat, with twenty-eight members operating fifty Earth stations. But "global" did not mean that all parts of the world would be treated equally.

Unlike the International Telecommunication Union, which traditionally relied on voting members represented by government officials and where each member had one vote, the Intelsat consortium apportioned votes based on investment share, which was calculated from a country's use of worldwide telecommunications traffic. Because the United States largely controlled the new Cold War technology that married rockets to communications satellites and was the major user of international communications, it dominated the voting. And because the United States dominated the negotiations for an organizational standard for global satellite communications (the arrangements with the interim agreement in 1964), it also played a major role in determining the technical standards for the Intelsat system, including the decision in the mid-1960s to adopt geosynchronous satellites, which would appear to an observer on the ground to be stationary, instead of medium-altitude satellites that would need to be tracked by ground stations.[1] Because the Earth stations used with geosynchronous satellites were less complex and therefore cheaper than the ground facilities with medium-altitude satellites, this decision was consistent with the US objective of maximizing the participation of poorer countries in the Global South.

The introduction of the first global satellite communications system supported the long-term trend identified in the beginning of the book—the ascendance of the United States in the history of international communications. The country started to challenge British dominance beginning in the 1920s; with this new technology, we see the completion of the process during the post–World War II period. In the Cold War context of nationalist globalism, however, the United States was committed to prioritizing a global infrastructure potentially benefiting not only the United States but also poorer countries around the world. The Soviet Union was the only country that might have provided an alternative to Intelsat during the 1960s and 1970s. The Soviets did establish a competing Intersputnik system in 1971, but it was much more limited than Intelsat, primarily serving the communist countries allied with the Soviet Union.[2]

Negotiations among all Intelsat members for definitive agreements, which came into force in 1973, began in 1969. The United States thus succeeded in using the interim arrangements to dominate the global system for nearly a decade. And the US government continued to make key decisions during the 1960s and 1970s, which helps explain how the system successfully globalized, including a decision by President Lyndon Johnson, in March 1966, directing government agencies to support the establishment of ground stations in poorer non-Western countries. National Security Action Memorandum No. 342 directed government officials to "take active steps to encourage the construction of earth-station links to the worldwide communication satellite system in selected less-developed countries." Specifically, the president ordered the State Department and the Agency for International Development (AID) to study the situation to make recommendations about providing direct support to specific countries. He stipulated that AID funds should be made available to countries wanting to establish ground stations but unable to obtain other sources of support.[3] Government officials were also instructed to assist countries in obtaining funding from international development banks or from private sources, including US and European telecommunications carriers.[4]

The United States succeeded in stimulating enthusiasm for its space achievements, and this was especially clear in developing countries with the introduction of satellite communications. Although developing countries did not participate in the initial negotiations, many nations in the Global South joined during the late 1960s and especially the 1970s. As early as September 1964, within a few weeks of the signing of the Intelsat initial agreements, a State Department official involved in trying to connect one region in Africa to the satellite system wrote that his work had "already been so publicized and has generated such interest amongst the Africans . . . that their exclusion from the project even in the early phases could have unfavorable political implications."[5] The official inauguration of ground stations in these countries often functioned as "satellite spectaculars," with national celebrations overseen by top political leaders. When Kenya held a special ceremony, in November 1970, announcing the opening of its first Intelsat Earth station, the local newspaper, the *East African Standard*, published a series of major articles about the event and included a picture of Kenyan president Jomo Kenyatta talking to the UN

Secretary General on the first official telephone call from the station. The front-page headline announced triumphantly, "Kenya Enters the Space Age."[6] Intelsat Earth stations not only served as symbols of US success in the Cold War competition with the Soviet Union to win over the hearts and minds of citizens in non-Western countries, but also had important implications for countries wanting to break free from colonial relationships. Leaders in many non-Western countries were receptive to US arguments stressing that the new space-age invention would allow global communications to "flow free of artificial constraints held over from the colonial traditions of past centuries."[7]

Although the agreements establishing Intelsat coordinated both organizational and technical standards for global satellite communications, they did not attempt to regulate the content of communications. This is consistent with a general pattern in global communications governance. National governments have actively sought to regulate the content of communications, but global governance has consistently focused on coordinating technical and organizational standards separate from regulating content. France raised concerns related to content during the debates leading to the 1964 interim agreements when the country's representatives spoke out against possibly using communications satellites for direct propaganda broadcasts to individual homes. The agreements establishing Intelsat, however, did not attempt to ban propaganda broadcasts from satellites.

The first direct broadcast satellites (DBS), transmitting directly to home satellite dishes, began experimental operations during the mid-1970s. One of the earliest efforts was the Satellite Instructional Television Experiment (SITE) in India using NASA's experimental *ATS*-6 satellite and special television receivers located in rural villages. SITE tested the benefits of direct satellite broadcasting for development efforts in a poor region of the world. More generally, the first regular DBS service began during the following decade, and commercial operations expanded significantly during the 1990s.[8]

In this book, I analyze the development of global infrastructure rather than communications and media content, mainly because of the focus on the formative period before Intelsat began major operations but also because communications infrastructure in general has been an understudied topic. Nonetheless, general comments about the implications for media content of the decisions analyzed in this book for the period after the establishment of Intelsat are appropriate. Had the Kennedy administration not overturned the previous policy of the Eisenhower administration, AT&T likely would have taken the lead in setting up a limited satellite communications system mainly serving wealthy countries in the Northern Hemisphere, especially during the late 1960s and 1970s. Although the Kennedy administration's decision to create a more inclusive global system opened up new possibilities for all countries during a period of "communications scarcity across the globe," it also supported the spread of American popular culture around the world, not only to countries in the Northern Hemisphere, particularly Western Europe, but also to countries in the Global South.[9] US media companies increasingly produced television programs and films for a global export market during this period, and the Intelsat system supported the distribution of these cultural products to countries that a more commercially oriented satellite communications system mainly interested in wealthy markets in the Northern Hemisphere would not have served.[10] In 1982, for example, US television programs made up 36 percent of total daily programming on Brazilian television.[11] Intelsat was also important for delivering live television broadcasts to a global audience; the broadcast of such major global "satellite spectaculars" highlighting American achievements, such as the 1969 moon landing, dramatically underscored the ascendance of the United States as a dominant world power. The Intelsat system, especially because of its global reach, thus supported US soft-power strategies that used cultural media products to expand US global influence.

But Intelsat's global influence arguably also helped inspire an important counter-movement during the 1970s, mainly among nonaligned nations at the UN, who advocated for a new world information and communication order aimed particularly at replacing US "media imperialism," a specific type of cultural imperialism that, as expressed by an influential critic from Latin America, was based on the use of media as "the vehicle for transmitting values and lifestyles to Third World countries which stimulate the type of consumption and the type of society necessary to the transnational system as a whole."[12] The movement grew from a broader proposal from the Non-Aligned Movement for a New International Economic Order. In 1974 these nations convinced the UN General Assembly to adopt several fundamental principles for a new economic order, including a "progressive social transformation that enables the full participation of the population in the development process."[13] Support from UNESCO and the UN General Assembly led to the establishment of a special UNESCO commission—called the McBride Commission after the chair, Sean McBride, a diplomat from Ireland who won the 1974 Nobel Peace Prize. The commission issued a famous report in 1980 called *Many Voices, One World: Towards a New, More Just, and More Efficient World*

Information and Communication Order.[14] UNESCO then formally adopted the general recommendations of the report at a meeting of the General Assembly in Belgrade that same year.[15]

The proposed new communication order would result in more balanced communication flows between different regions of the world, replacing a prevailing one-way flow from the Global North, especially from the United States and its overpowering global media conglomerates. The report was concerned not only about imbalances in the flow of entertainment-oriented television and film programs but also about the global flow of news between different regions. Critics advocated replacing distorted, and generally negative, Western news stories about the Global South with balanced reporting that respected "the rights of each nation to inform the world public about its interests, its aspirations and its social and cultural values."[16] Finally, the report also recommended eliminating media and communication inequalities to address the dependence of the Global South on wealthy countries in the Global North, including dependence on information-technology hardware and software as well as on all types of communication technologies.

In some cases, poorer countries in the Global South arguably would have been better off investing limited resources in basic telecommunications infrastructure for local needs rather than purchasing expensive Intelsat-linked Earth stations. And international point-to-point communication using the Intelsat system—telephony, telex, facsimile, and so forth—especially benefited local elites with ties to multinational corporations. Through these uses, the Intelsat system also worked to bind Global South countries to the United States and other wealthy countries in the Northern Hemisphere.

The case of the first country in Africa to set up an Earth station, Morocco, indicates that at least some developing countries could not justify the investment based on existing needs but were mainly motivated by a desire to cooperate with the United States. The Radio Corporation of America (RCA) initially had declined a request from the Moroccan government to finance and build the station, arguing that it "would not yet be profitable in Morocco."[17] Although other support was found to build the station, a month after the start of operations, in January 1970, the government discovered that the venture was not viable due to lack of business. Only three channels were being used. To break even, the station needed to operate twenty-four channels.[18] A Moroccan government official privately admitted that the country "had undertaken [the] satellite communications venture more because of [a] desire to cooperate with the US Government in expanding [the] satellite communications network than because of [a] real need for [an] earth station in Morocco." He insisted that the United States had a commitment to help by ordering the two major US organizations operating in Morocco, the US Navy and the Voice of America, to use the facilities. According to this same official, the failure of the venture in Morocco "to become viable because of lack of business would reflect adversely on US technology and retard expansion of communications by satellite."[19] A staff member at the US embassy in Morocco agreed, arguing, "I am sure there is no doubt in anyone's mind that it is in the US interest that this first satellite communications station in Africa be a success."[20]

In general, although the influence of the United States in the operation of Intelsat was still significant following the signing of the definitive agreements, other countries did become more important. New members that joined the consortium after 1973 mainly represented developing countries. By the 1990s, they represented more than 80 percent of members, and the organization emphasized its special commitment to these members: "It is Intelsat's special duty and obligation to promote the cause of communication development that distinguishes Intelsat in a substantive and important way."[21]

Because voting by the most important body established by the definitive agreements, the board of governors, was based on usage, however, the largest number of votes went mainly to wealthy developed countries. In the mid-1990s, the five largest users of Intelsat—the United States, the United Kingdom, Japan, France, and Germany—controlled more than 40 percent of investment shares. OECD countries, which mainly represented "high-income" or "developed" countries, controlled more than 60 percent of shares. Shares controlled by individual developing countries were small by comparison (around 2 percent each for China and India, for example), and their separate interests kept them from acting as a bloc.[22]

Intelsat's emphasis on serving developing countries during the 1990s was not only a legacy of the 1960s global Cold War context but also a business strategy in response to new competition. The first competition came from satellites launched to serve domestic or regional markets. Intelsat was originally established to provide international satellite communications, or, in other words, satellite links across national borders. But in the 1970s it began to offer domestic service to countries that did not have alternative terrestrial facilities, mainly developing countries in Asia, Latin America, and Africa. The first country served was Algeria in 1976. By the mid-1990s more than twenty developing countries used Intelsat satellites for domestic service, primarily telephone and data links between major cities and the national distribution of television programming.[23] Beginning in the early 1980s, Intelsat sponsored an innovative development program called Project SHARE, which used one of its satellites to provide rural communities in third world countries with educational and medical assistance.[24]

Intelsat did not have a monopoly in domestic satellite communications (unlike its monopoly in international satellite communications), and the first national satellites provided alternatives to the domestic service offered by Intelsat. Canada was the first country to own a geostationary satellite used for domestic communications (serving especially the country's vast northern regions). NASA launched *Anik A* in 1972; Hughes built the satellite, but important subcontracts went to Canadian manufacturers. The first US domestic communications satellite, *WESTAR 1* (built by Hughes for Western Union), was launched in 1974. The major service initially for domestic satellites in the United States was pay television. HBO (Home Box Office) used RCA's *Satcom 1*, launched in 1976, to transmit to domestic cable systems, and this satellite connection played an important role in the growth of cable television in the United States. In 1976, Indonesia became the first developing country to have its own domestic communications satellite, *Palapa A*, built by Hughes and launched by NASA. India gained its first operational communications satellite in 1982 (also launched by NASA but built by Ford Aerospace). China manufactured and launched its first geostationary communications satellite in 1984. Hughes built Mexico's first domestic communications satellite, and NASA launched it in 1985. Brazil used Hughes as a secondary contractor and Spar Aerospace of Canada as a primary contractor for its first communications satellite, but the European Space Agency launched it (also in 1985).[25] Pakistan built its own domestic communications satellite in 1990 but used a Chinese launcher.[26]

Intelsat faced more substantial competition beginning during the 1980s from communications satellites serving regional markets. The 1971 Intelsat definitive agreements did not exclude members from establishing separate regional systems, but article XIV(d) stipulated that members planning a separate system would need to first demonstrate that it would not cause "significant economic harm" to the Intelsat global system.[27] As we have seen, countries in Europe had long desired to create separate systems, and with the European Space Agency's development of the Ariane launch vehicle, a separate system serving Western Europe became possible. The European Conference of Postal and Telecommunications Administrations (CEPT) provisionally established the European Telecommunications Satellite Organization (Eutelsat) in 1977.

Eutelsat then successfully convinced the Intelsat Board of Governors and Assembly of Parties that by observing specific restrictions, it would not substantially compete with the global consortium. The first Eutelsat satellite was launched in 1983.

The decision during the mid-1970s to establish a separate international organization serving maritime satellite communications, INMARSAT, likely provided a precedent for Eutelsat, which, in turn, opened the door for other regional systems.[28] Arabsat (the Arab Satellite Communications Organization), a regional system providing coverage for Arab countries in the Middle East and North Africa, orbited its first satellite in 1985, and AsiaSat (Asia Telecommunications Holdings Limited), based in China but serving most of Asia, launched its first satellite in 1990. By providing service to the company StarTV, AsiaSat became especially important in the spread of international television across Asia. By the mid-1990s, more than 200 million viewers in fifty-plus countries had access to new television programming not available from their domestic television systems.[29]

A much more substantial threat to Intelsat's global monopoly came from private US satellites. Supported by the ascendance of neoliberal values and market-centered policies during the 1980s and 1990s, private satellite communications companies sought to introduce competition by undermining state-centered control of markets. Similar efforts outside the United States led to the restructuring and privatization of national PTTs (post, telegraph, and telephone agencies) around the world, based on market-centered frameworks. The recommendations of the McBride Commission for a new world information and communication order died during this period in the face of opposition from Western countries, led by the Ronald Reagan administration in the United States, which withdrew from UNESCO in protest of its activities in 1984. The United States favored a "free flow of information" and argued that the call for a new communications order mainly represented a movement to control media through state regulation and government censorship. The Telecommunications Act of 1996 led to major changes in the organization of communications in the United States based on policies of deregulation and liberalization. And the ITU also adapted to this new era and began to champion neoliberal values.[30]

In 1983, RCA and two new companies, PanAmSat and Orion, applied to the US government to compete directly with Intelsat by operating private communications satellite systems serving international communications. The application from Orion was motivated especially by an ideological commitment to deregulation. Key leaders in the company had served as congressional aides allied with members of the Reagan administration, who were committed to breaking up AT&T and other regulated monopolies. Television broadcasters critical of the amount Intelsat charged for international television distribution led PanAmSat.[31]

The Reagan administration generally supported these new initiatives. In November 1984, President Reagan signed a declaration affirming that alternative satellite systems were "required in the national interest." The next year, the FCC issued a "Separate Systems decision," which declared that "separate systems will provide substantial benefits to the users of international satellite communications services without causing significant economic harm to Intelsat."[32] Although the FCC decision technically authorized five companies to attempt to compete with Intelsat, it also established difficult entry requirements that limited the initial impact of the decision. Nevertheless, first PanAmSat and then other private companies competed with Intelsat. PanAmSat launched its first satellite in 1988, focusing initially on markets in Latin America, the Caribbean, and the southern regions of the United States, as well as in the Atlantic region in general.[33] The company took advantage especially of the growing market for international television; CNN was the company's first customer. Orion was slower to develop, not launching its first satellite until 1994.[34]

Private satellites did not represent the only major challenge for Intelsat; a new type of submarine cable, fiber-optic cable, challenged not only Intelsat's core business but also the general dominance of satellites in international communications. The first transatlantic fiber-optic cable, TAT-8, started service in 1988. It could handle forty thousand voice conversations. TAT-9 and TAT-10, both introduced in 1992, doubled the capacity of TAT-8. TAT-14, which began operating in 2001, had a capacity equivalent to an astounding 9.7 million telephone channels.[35] These improvements and the dramatic expansion of undersea fiber-optic cables around the world during this period were also driven by privatization in the industry. While TAT-8 continued the earlier tradition of having undersea cables built and owned by AT&T and European telecommunications administrations, most later cables were built and owned by private companies, which, in many cases, did not have previous experience in international communications.[36]

Since the 1990s, while fiber-optic undersea cable networks have increasingly handled international voice, data, and internet traffic, new generations of Intelsat satellites have also developed better communications capacity and speed. For example, Intelsat's seventh generation satellite, *Intelsat 701*, launched in 1993, could handle approximately ninety thousand telephone calls.[37] Yet, fiber-optic cables hold advantages relative to satellites that do not involve capacity or speed. Most important, cables do not suffer from the signal delay resulting from the large distances transmissions must travel between Earth stations and geosynchronous satellites (potentially as high as a quarter of a second for each route). They also are more secure and not prone to atmospheric interference or disruption from extreme solar activity.[38] Because of these advantages and because they proved to be cheaper overall, by 2003, according to one estimate, fiber-optic cables had "94.4 percent of the world's transmissions capacity (compared to only 16 percent in 1988)."[39]

But, as with radio and cables in earlier eras, the two competing technologies learned to coexist and complement one another, partly by carving out separate markets. Communications satellites largely ceded point-to-point voice communications to new fiber-optic cables and focused on specialized services, particularly video transmissions, especially to regions not served by cables.[40] To compete with fiber-optic cables during the 1990s, Intelsat offered new services, expanded its capacity for domestic service, and lowered tariffs.[41] To understand the resilience of cables, it is also important to place them in a historical context that recognizes their unique strategic value for the United States and other countries. As Jonathan Winkler has argued, "The military and governmental needs for the secrecy inherent in cable-based communications ensures that the strategic importance of global cable networks first envisioned by Great Britain before 1914 continues to be of chief importance for the United States in the next century."[42]

In 2001, after years of having its protected status weakened, Intelsat was privatized. Wealthy countries generally pushed the hardest for this result. Developing countries were less enthusiastic; they feared that a private, commercial organization would not have any reason to serve their needs.[43] This was an important initial reason for setting up the Intelsat global consortium. US officials had worried that if they did not act, AT&T would have dominated satellite communications, and because it was primarily a commercial company, it would not have had an incentive to create a true global system open to serving all countries. To satisfy public service obligations such as this fundamental commitment to serving developing countries, a separate intergovernmental institution was created at the same time Intelsat became a private company: the International Telecommunications Satellite Organization (ITSO).[44] ITSO officially incorporated the basic principle established in 1961 with Resolution 1721 (XVI) of the General Assembly of the United Nations, that "communication by means of satellites should be available to the nations of the world as soon as practicable on a global and non-discriminatory basis." In 2020, ITSO's official mission was "to monitor the performance of Intelsat's public service obligations and to safeguard the Parties' Common Heritage."[45]

The privatization and commercialization of satellite communications has not meant the end of global concerns. Since the 1980s, a global focus has persisted, but it has been redefined. As Martin Collins has argued, the end of the Cold War led to a new understanding of "global" that was no longer state centered but based on the central role of markets and private corporations. According to Collins, private satellite communications companies first established during the 1990s—including PanAmSat, Iridium, and Globalstar— "gave the global a powerful new connotation, not just as a metaphor for transnational flows of money, processes of production, and culture, but as an indicator of an emergent capability to project market values, power, and control over the totality of the planet."[46]

Since their introduction in the late 1950s, communications satellites have played a key role in driving globalization processes and in how we think about the global. And Collins's observation reminds us that this role has changed during different eras. The first global satellite communications system was very much a product of the global Cold War, in particular the US Cold War commitment to nationalist globalism. And the specific notion of global connected with Intelsat also reflected specific characteristics connected with this era. In particular, the United States used the notion of global to win over hearts and minds through soft power and propaganda, to promote development tools and assistance strategies, to champion spectacular world-spanning technological innovations, to shape decisions about technical standards at international institutions, and in general to project state-based political, economic, and military power around the world.

Introduction

1. "Apollo 11 Mission Overview," NASA, last updated May 15, 2019, https://www.nasa.gov /mission_pages/apollo/missions/apollo11.html. Also see David Meerman Scott and Richard Jurek, *Marketing the Moon: The Selling of the Apollo Lunar Program* (Cambridge, Mass.: MIT Press, 2014), xi–xii.

2. Lisa Parks, *Cultures in Orbit: Satellites and the Televisual* (Durham, N.C.: Duke University Press, 2005).

3. Michael Allen, *Live from the Moon: Film, Television and the Space Race* (London: I. B. Tauris, 2009), 145, 149–53; James Schwoch, *Global TV: New Media and the Cold War, 1946–69* (Urbana: University of Illinois Press, 2009), 1–2.

4. Allen, *Live from the Moon*, 56–59; Schwoch, *Global TV*, 128; James Hay, "The Invention of Air Space, Outer Space, and Cyberspace," in *Down to Earth: Satellite Technologies, Industries, and Cultures*, ed. Lisa Parks and James Schwoch (New Brunswick, N.J.: Rutgers University Press, 2012), 31; Aniko Bodroghkozy, "The BBC and the Black Weekend: Broadcasting the Kennedy Assassination and the Birth of Global Television News," *The Sixties: A Journal of History, Politics and Culture* 9 (2016): 242–60.

5. Christy Collis, "The Geostationary Orbit: A Critical Legal Geography of Space's Most Valuable Real Estate," in Parks and Schwoch, *Down to Earth*, 61–81.

6. "Our World on Your Screen," CSIRO, accessed April 13, 2020, https://www.parkes.atnf .csiro.au/people/sar049/our_world/PDF/Our_World_on_Your_Screen_OCR.pdf.

7. Marshall McLuhan and Bruce R. Powers, *The Global Village: Transformations in World Life and Media in the 21st Century* (New York: Oxford University Press, 1992).

8. The fullest discussion of the *Our World* broadcast is Lisa Parks, "Our World, Satellite Televisuality, and the Fantasy of Global Presence," in *Planet TV: A Global Television Reader*, ed. Lisa Parks and Shanti Kumar (New York: New York University Press, 2004), 74–93. Also see discussion of a 1965 international television event called *Town Meeting of the World* in Kevin Grieves, "'A New Age of Diplomacy': International Satellite Television and *Town Meeting of the World*," *Journalism History* 40 (2014): 98–107.

9. This study is the first to provide a detailed history of Comsat and Intelsat using wide-ranging archival and published records focusing especially on the role of the United States in dominating the first global satellite communications system. For the best sources otherwise, see Schwoch, *Global TV*, 139–55; and David Whalen, *The Origins of Satellite Communications, 1945–1965* (Washington, D.C.: Smithsonian Institution Press, 2002), which focuses on the role of private enterprise, in particular AT&T and Hughes, in the early development of satellite communications in the United States. In contrast to Whalen's 2002 book, this work uses a wide

range of archival sources, especially from government archives in different countries, to understand the origins of the Intelsat global system. Also see David Whalen "Billion Dollar Technology: A Short Historical Overview of the Origins of Communications Satellite Technology, 1945–1965," in *Beyond the Ionosphere: Fifty Years of Satellite Communication*, ed. Andrew J. Butrica (Washington, D.C.: NASA History Office, 1997); and chapter 1 of David Whalen, *The Rise and Fall of Comsat: Technology, Business, and Government in Satellite Communications* (Basingstoke: Palgrave Macmillan, 2014). Whalen's 2014 book on Comsat explores the reasons for the decline of the company during the last decades of the twentieth century and its dissolution in 2001. Nonhistorians have also written about Intelsat. See Jonathan F. Galloway, *The Politics and Technology of Satellite Communications* (Lexington, Mass.: Lexington Books, 1972); Michael E. Kinsley, *Outer Space and Inner Sanctums: Government, Business, and Satellite Communication* (New York: Wiley, 1976); Delbert D. Smith, *Communication via Satellite: A Vision of Retrospect* (Boston: Sijthoff, 1976); Roger A. Kvam, "COMSAT: The Inevitable Anomaly," in *Knowledge and Power: Essays on Science and Government*, ed. Sanford A. Lakoff (New York: Free Press, 1966), 271–92; Marcellus S. Snow, *International Commercial Satellite Communications: Economic and Political Issues of the First Decade of INTELSAT* (New York: Praeger, 1987); Judith Teggar Kildow, *INTELSAT: Policymaker's Dilemma* (Lexington, Mass.: Lexington Books, 1973); Joel Alper and Joseph N. Pelton, eds. *The INTELSAT Global Satellite System* (New York: American Institute of Aeronautics and Astronautics, 1984); and Joseph N. Pelton, *Global Communications Satellite Policy: INTELSAT, Politics, and Functionalism* (Mt. Airy, Md.: Lomond Books, 1974).

10. For important insights from infrastructure studies, see Paul N. Edwards, "Introduction: An Agenda for Infrastructure Studies," *Journal of the Association for Information Systems* 10 (May 2009): 364–74; Susan Leigh Star, "The Ethnography of Infrastructure," *American Behavior Scientist* 43 (November/December 1999): 377–91; Andreas Marklund and Mogens Rüdiger, eds., *Historicizing Infrastructure* (Aalborg, Denmark: Aalborg Universitetsforlag, 2016); and Paul N. Edwards, Steven J. Jackson, Geoffrey C. Bowker, and Cory P. Knobel, *Understanding Infrastructure: Dynamics, Tensions, and Design—Report of a Workshop on "History & Theory of Infrastructure: Lessons for New Scientific Cyberinfrastructure,"* January 2007, https://deepblue.lib.umich.edu/bitstream/handle/2027.42/49353/UnderstandingInfrastructure2007.pdf?sequence=3&isAllowed=y.

11. On the history of communications infrastructure in a global context, see especially Schwoch, *Global TV*; James Schwoch, *The American Radio Industry and Its Latin American Activities, 1900–1939* (Urbana: University of Illinois Press, 1990); Simon M. Müller, *Wiring the World: The Social and Cultural Creation of Global Telegraph Networks* (New York: Columbia University Press, 2016); Nicole Starosielski, *The Undersea Network* (Durham, N.C.: Duke University Press, 2015); Heidi J. S. Tworek, *News from Germany: The Competition to Control World Communications, 1900–1945* (Cambridge, Mass.: Harvard University Press, 2019); Jonathan Reed Winkler, *Nexus: Strategic Communications and American Security in World War I* (Cambridge, Mass.: Harvard University Press, 2008); Jonas Brendebach, Martin Herzer, and Heidi Tworek, eds., *International Organizations and the Media in the Nineteenth and Twentieth Centuries: Exorbitant Expectations* (London: Routledge, 2018); Gabriele Balbi and Andreas Fickers, eds., *History of the International Telecommunication Union: Transnational Techno-Diplomacy from the Telegraph to the Internet* (Berlin: De Gruyter Oldenbourg, 2020); Daniel R. Headrick, *The Invisible Weapon: Telecommunications and International Politics, 1851–1945* (Oxford: Oxford University Press, 1991); and Andreas Fickers and Pascal Griset, *Communicating Europe: Technologies, Information, Events* (Berlin: Springer, 2018). A book on the history of transportation infrastructure in a global context that is especially important for this study is Jenifer Van Vleck, *Empire of the Air: Aviation and The American Ascendancy* (Cambridge, Mass.: Harvard University Press, 2013).

12. Star, "Ethnography of Infrastructure," 380.

13. Starosielski, *Undersea Network*, 3–6.

14. See especially Thomas Parke Hughes, *Networks of Power: Electrification in Western Society, 1880–1930* (Baltimore, Md.: Johns Hopkins University Press, 1983); Wiebe Bijker, Thomas P. Hughes, and Trevor Pinch, eds., *The Social Construction of Technological Systems* (Cambridge, Mass.: MIT Press, 1987); Wiebe E. Bijker and John Law, eds., *Shaping Technology/Building Society: Studies in Sociotechnnical Change* (Cambridge, Mass.: MIT Press, 1992); Oliver Cautard, *The Governance of Large Technical Systems* (New York: Routledge, 1999); Renate Mayntz and Thomas P. Hughes, *The Development of Large Technical Systems* (Boulder, Colo.: Westview Press, 1988): Thomas P. Hughes, *American Genesis: A Century of Invention and Technological Enthusiasm, 1870–1970* (Chicago: University of Chicago Press, 2004); and Thomas P. Hughes, *Rescuing Prometheus: Four Monumental Projects that Changed Our World* (New York: Vintage, 2011). For additional discussion about the value of analyzing communication projects in terms of large-scale technical systems, see Gabriele Balbi and Richard John, "Point-to-Point: Telecommunications Networks from the Optical Telegraph to the Mobile Telephone," in *Communication and Technology*, ed. Lorenzo Cantoni and James A. Danowski (Berlin: De Gruyter Mouton, 2015), 35–55, on 45.

15. Edwards et al., *Understanding Infrastructure*, 3.

16. JoAnne Yates and Craig N. Murphy, *Engineering Rules: Global Standard Setting since 1880* (Baltimore, Md.: Johns Hopkins University Press, 2019), 2. Yates and Murphy's book provides the best general overview of the history of standards and standardization; however, unlike this book, in which I explore the US government's role in the development of international standards for radio frequency allocation, government or intergovernmental standards setting was not their focus. Yates and Murphy are primarily interested in the private, nongovernmental bodies involved in standards setting. Most other studies in the history of standards focus on specific institutions or specific industries. Some of the more important books include Andrew L. Russell, *Open Standards and the Digital Age: History, Ideology and Networks* (Cambridge: Cambridge University Press, 2014); Shane Greenstein, *How the Internet Became Commercial: Innovation, Privatization, and the Birth of a New Network* (Princeton, N.J.: Princeton University Press, 2015); Hugh Richard Slotten, *Radio and Television Regulation: Broadcast Technology in the United States, 1920–1960* (Baltimore, Md.: Johns Hopkins University Press, 2000); Craig N. Murphy and JoAnne Yates, *The International Organization for Standardization: Global Governance through Voluntary Consensus* (London: Routledge, 2009); Martha Lampland and Susan Leigh Star, eds., *Standards and Their Stories: How Quantifying, Classifying, and Formalizing Practices Shape Everyday Life* (Ithaca, N.Y.: Cornell University Press, 2008); Balbi and Fickers, *History of the International Telecommunication Union*; Lawrence Busch, *Standards: Recipes for Reality* (Cambridge, Mass.: MIT Press, 2010); and Steven W. Usselman, *Regulating Railroad Innovation: Business, Technology, and Politics in America, 1840–1920* (Cambridge: Cambridge University Press, 2002).

17. Edwards et al., *Understanding Infrastructure*, 24.

18. Immerwahr quoted by Timothy Nunan, "Thinking Big . . . and Small about U.S. History in a Global Context with Daniel Immerwahr," Toynbee Prize Foundation, February 23, 2015, https://toynbeeprize.org/posts/daniel-immerwahr/.

19. Headrick, *Invisible Weapon*, 38

20. Müller, *Wiring the World*, 52. The literature on the history of the electric telegraph is extensive. For works emphasizing the importance of undersea cables, see especially Headrick, *Invisible Weapon*; Winkler, *Nexus*; Dwayne R. Winseck and Robert M. Pike, *Communication and Empire: Media, Markets, and Globalization, 1860–1930* (Durham, N.C.: Duke University Press, 2007); Tworek, *News from Germany*; Bernard Finn and Daqing Yang, eds., *Communications Under the Seas: The Evolving Cable Network and Its Implications* (Cambridge, Mass.: MIT Press, 2009); Daqing Yang, *Technology of Empire: Telecommunication and Japanese Expansion in Asia,*

1883-1945 (Cambridge, Mass.: Harvard University Press, 2010); Jonathan Silberstein-Loeb, *The International Distribution of News: The Associated Press, Press Association, and Reuters, 1848-1947* (Cambridge: Cambridge University Press, 2014); John A. Britton, *Cables, Crises, and the Press: The Geopolitics of the New International Information System in the Americas, 1866-1903* (Albuquerque: University of New Mexico Press, 2013); Jill Hills, *Telecommunications and Empire* (Urbana: University of Illinois Press, 2007); David Paull Nickles, *Under the Wire: How the Telegraph Changed Diplomacy* (Cambridge, Mass.: Harvard University Press, 2003); Ken Beauchamp, *History of Telegraphy* (London: Institute of Electrical Engineers, 2001); Roland Wenzlhuemer, *Connecting the Nineteenth-Century World: The Telegraph and Globalization* (Cambridge: Cambridge University Press, 2013); Simon J. Potter, *News and the British World: The Emergence of an Imperial Press System, 1876-1922* (Oxford: Oxford University Press, 2005); Peter J. Hugill, *Global Communications since 1844: Geopolitics and Technology* (Baltimore, Md.: Johns Hopkins University Press, 1999); Michaela Hampf and Simone Müller-Pohl, eds. *Global Communication Electric: Business, News and Politics in the World of Telegraphy* (Frankfurt: Campus, 2013); and Amelia Bonea, *The News of Empire: Telegraphy, Journalism, and the Politics of Reporting in Colonial India, c. 1830-1900* (Oxford: Oxford University Press, 2016).

21. Winkler, *Nexus*, 13-14.

22. Scholars such as Winseck and Pike have emphasized that this control should not be overemphasized. The Eastern Telegraph Company and other multinational cable companies "were less closely allied with national capital and governments than commonly assumed." See Winseck and Pike, *Communication and Empire*, xvi. Also see Tworek, *News from Germany*, 48.

23. Headrick, *Invisible Weapon*, 88-9; Tworek, *News from Germany*, 48-49.

24. Winseck and Pike, *Communications and Empire*, 7; Headrick, *Invisible Weapon*, 50-72.

25. Tworek, *News from Germany*, 48-49; Headrick, *Invisible Weapon*, 102-5.

26. Robert Boyce, "Imperial Dreams and National Realities: Britain, Canada and the Struggle for a Pacific Telegraph Cable, 1879-1902," *English Historical Review* 115 (2000): 39-70.

27. Headrick, *Invisible Weapon*, 195-96.

28. Headrick, *Invisible Weapon*, 130-33.

29. Tworek, *News from Germany*, 49; Headrick, *Invisible Weapon*, 121-30.

30. Headrick, *Invisible Weapon*, 173-80; postmaster general quoted on 175.

31. Headrick, *Invisible Weapon*, 125-27, 180-83. On the history of point-to-point radio communication involving the United States, see Hugh G. J. Aitken, *The Continuous Wave: Technology and American Radio, 1900-1932* (Princeton, N.J.: Princeton University Press, 1985); Susan J. Douglas, *Inventing American Broadcasting, 1899-1922* (Baltimore, Md.: Johns Hopkins University Press, 1987); Slotten, *Radio and Television Regulation*; Schwoch, *American Radio Industry*; Michael A. Krysko, *American Radio in China: International Encounters with Technology and Communications, 1919-41* (Basingstoke: Palgrave Macmillan, 2011); and Winkler, *Nexus*.

32. Headrick, *Invisible Weapon*, 202-5.

33. Jonathan Reed Winkler, "Bridging the Gap: The Cable and Its Challengers, 1919-1988," in Finn and Yang, *Communications Under the Seas*, 30-31.

34. Headrick, *Invisible Weapon*, 205-8. In Yang's study of telecommunications and Japanese expansion in Asia, he argues that the formation of Cable and Wireless provided a model for Japan. In response to the similar challenges presented by radio, the government "was able to modify the strict state monopoly and create close state-business collaborations without giving up total control." See Yang, *Technology of Empire*, 403.

35. Winkler, "Bridging the Gap," 33-35; Headrick, *Invisible Weapon*, 208-10. On merger efforts, also see Winseck and Pike, *Communication and Empire*, 328-29.

36. Winkler, "Bridging the Gap," 36-39.

37. Schwoch, *American Radio Industry*, 96-123.

38. Headrick, *Invisible Weapon*, 267. Also see Brian McKercher, *Transition of Power: Britain's*

Loss of Global Pre-eminence to the United States, 1930–1945 (New York: Columbia University Press, 1999).

39. Daniel Headrick, "Public-Private Relations in International Telecommunications before World War II," in *Telecommunications Politics: Ownership and Control of the Informational Highway in Developing Countries*, ed. Bella Mody, Johannes M. Bauer, and Joseph D. Staubhaar (Mahwah, N.J.: Erlbaum, 1994). Also see Balbi and John, "Point-to-Point," 46.

40. Headrick, *Invisible Weapon*, 190. On US government support for radio companies operating in Latin America, also see Schwoch, *American Radio Industry*.

41. On the implications of Cold War military research, see, for example, Naomi Oreskes and John Krige, eds., *Science and Technology in the Global Cold War* (Cambridge, Mass.: MIT Press, 2014); David Hounshell, "The Cold War, RAND, and the Generation of Knowledge," *Historical Studies in the Physical and Biological Sciences* 27 (1997): 237–67; Daniel J. Kevles, "Cold War and Hot Physics: Science, Security, and the American State, 1945–56," *Historical Studies in the Physical and Biological Sciences* 20 (1990): 239–64; Stuart W. Leslie, *The Cold War and American Science: The Military-Industrial-Academic Complex at MIT and Stanford* (New York: Columbia University Press, 1993); Martin J. Collins, *Cold War Laboratory: RAND, the Air Force, and the American State, 1945–1950* (Washington, D.C.: Smithsonian Institution Press, 2002); Rebecca S. Lowen, *Creating the Cold War University: The Transformation of Stanford* (Berkeley: University of California Press, 1997); David H. DeVorkin, *Science with a Vengeance: How the Military Created U.S. Space Science after World War II* (New York: Springer, 1992); Thomas J. Misa, "Military Needs, Commercial Realities, and the Development of the Transistor, 1948–1958," in *Military Enterprise and Technological Change: Perspectives on the American Experience*, ed. Merritt Roe Smith (Cambridge, Mass.: MIT Press, 1985), 253–88; Daniel Kevles, "K1S2: Korea, Science, and the State," in *Big Science: The Growth of Large-Scale Research*, ed. Peter Galison and Bruce Hevly (Stanford, Calif.: Stanford University Press, 1992), 312–33; Thomas J. Misa, "Command Performance: A Perspective on Military Enterprise and Technological Change," in Smith, *Military Enterprise and Technological Change*, 329–46; and Ronald E. Doel, Kristine C. Harper, and Matthias Heymann, eds., *Exploring Greenland: Cold War Science and Technology on Ice* (New York: Palgrave Macmillan, 2016). On the role of science and technology in weapons development and diplomacy, see especially Gregg Herken, *Cardinal Choices: Presidential Science Advising from the Atomic Bomb to SDI*, rev. ed. (Stanford, Calif.: Stanford University Press, 2000); Robert Gilpin and Christopher Wright, eds., *Scientists and National Policy-Making* (New York: Columbia University Press, 1964); Matthew Evangelista, *Unarmed Forces: The Transnational Movement to End the Cold War* (Ithaca, N.Y.: Cornell University Press, 1999); Harold Karan Jacobson and Eric Stein, *Diplomats, Scientists, and Politicians: The United States and the Nuclear Test Ban Negotiations* (Ann Arbor: University of Michigan Press, 1966); and Joseph Manzione, "'Amusing and Amazing and Practical and Military': The Legacy of Scientific Internationalism in American Foreign Policy, 1945–1963," *Diplomatic History* 24 (Winter 2000): 21–55. Also see the special issue of *Osiris* 21 (2006), including the following articles: Clark A. Miller, "An Effective Instrument of Peace: Scientific Cooperation as an Instrument of U.S. Foreign Policy, 1938–1950," 133–60; John Krige, "Atoms for Peace, Scientific Internationalism, and Scientific Intelligence," 161–81; Gabrielle Hecht, "Negotiating Global Nuclearities: Apartheid, Decolonization, and the Cold War in the Making of the IAEA," 25–48; and Kai-Henrik Barth, "Catalysts of Change: Scientists as Transnational Arms Control Advocates in the 1980s," 182–208. Also see Audra J. Wolfe, *Freedom's Laboratory: The Cold War Struggle for the Soul of Science* (Baltimore, Md.: Johns Hopkins University Press, 2018); and Paul Rubinson, *Redefining Science: Scientists, the National Security State, and Nuclear Weapons in Cold War America* (Amherst: University of Massachusetts Press, 2018).

42. See, for example, Melvyn P. Leffler, *For the Soul of Mankind: The United States, the Soviet Union, and the Cold War* (New York: Hill and Wang, 2007); Nicholas J. Cull, *The Cold War and*

the United States Information Agency: American Propaganda and Public Diplomacy, 1945–1989 (Cambridge: Cambridge University Press, 2008); John Krige, *American Hegemony and the Postwar Reconstruction of Science in Europe* (Cambridge, Mass: MIT Press, 2006); Walter Lafeber, "Technology and U.S. Foreign Relations," *Diplomatic History* 24 (2000): 1–19; Michael E. Latham, *Modernization as Ideology: American Social Science and the 'Nation Building' in the Kennedy Era* (Chicago: University of Chicago Press, 2000); David Reynolds, *One World Divisible: A Global History since 1945* (New York: Norton, 2000); and Odd Arne Westad, *The Global Cold War: Third World Interventions and the Making of Our Times* (New York: Cambridge University Press, 2007). The term "soft power" was first popularized by Joseph Nye. See his *Bound to Lead: The Changing Nature of American Power* (New York: Basic Books, 1990). Also see Joseph Nye, *Soft Power: The Means to Success in World Politics* (New York: Public Affairs, 2004).

43. On the international implications of space exploration during the Cold War, see especially John Krige, "Technology, Foreign Policy, and International Cooperation in Space," in *Critical Issues in the History of Spaceflight*, ed. Steven J. Dick and Roger D. Launius (Washington, D.C.: NASA, 2006); and Walter A. McDougall, . . . *The Heavens and the Earth: A Political History of the Space Age* (Baltimore, Md.: Johns Hopkins University Press, 1985). For other sources dealing with the interaction between science and technology and Cold War diplomacy that take a broader perspective, see Allan A. Needell, *Science, Cold War, and the American State: Lloyd V. Berkner and the Balance of Professional Ideals* (Amsterdam: Harwood Academic, 2000); Clark A. Miller, "Scientific Internationalism in American Foreign Policy: The Case of Meteorology, 1947–1958," in *Changing the Atmosphere: Expert Knowledge and Environmental Governance*, ed. Clark A. Miller and Paul N. Edwards (Cambridge, Mass.: MIT Press, 2001), 167–218; Nick Cullather, "Miracles of Modernization: The Green Revolution and the Apotheosis of Technology," *Diplomatic History* 28 (April 2004): 227–54; John H. Perkins, *Geopolitics and the Green Revolution: Wheat, Genes, and the Cold War* (Oxford: Oxford University Press, 1997); Jacob Hamblin, *Oceanographers and the Cold War: Disciples of Marine Science* (Seattle: University of Washington Press, 2005); Kurk Dorsey, "Dealing with the Dinosaur (and Its Swamp): Putting the Environment in Diplomatic History," *Diplomatic History* 29 (September 2005): 573–87; and Ronald E. Doel and Kristine C. Harper, "Prometheus Unleashed: Science as a Diplomatic Weapon in the Lyndon B. Johnson Administration," *Osiris* 21 (2006): 66–85. For an overview of the historical interaction among science, technology, and diplomacy, see especially John Krige and Kai-Henrik Barth, "Introduction: Science, Technology, and International Affairs," *Osiris* 21 (2006): 1–21; Ronald E. Doel and Zuoyue Wang, "Science and Technology," in *Encyclopedia of American Foreign Policy*, ed. Alexander DeConde, Richard Dean Burns, and Fredrik Logevall, rev. ed. (New York: Scribner, 2001), 443–59; Ronald E. Doel, "Scientists as Policymakers, Advisors, and Intelligence Agents: Linking Contemporary Diplomatic History with the History of Contemporary Science," in *The Historiography of Contemporary Science and Technology*, ed. Thomas Söderqvist (Amsterdam: Harwood Academic, 1997), 215–44; and Lafeber, "Technology and U.S. Foreign Relations," 1–19.

44. See especially Kenneth Osgood, *Total Cold War: Eisenhower's Secret Propaganda Battle at Home and Abroad* (Lawrence: University Press of Kansas, 2006). For the influence of the Cold War on social science in the United States, see Ronald W. Evans, *The Hope for American School Reform: The Cold War Pursuit of Inquiry Learning in Social Studies* (New York: Palgrave Macmillan, 2011); Nils Gilman, *Mandarins of the Future: Modernization Theory in Cold War America* (Baltimore, Md.: Johns Hopkins University Press, 2004); Mark Solovey, *Shaky Foundations: The Politics-Patronage-Social Science Nexus in Cold War America* (New Brunswick, N.J.: Rutgers University Press, 2013); and David Ekbladh, *The Great American Mission: Modernization and the Construction of an American World Order* (Princeton, N.J.: Princeton University Press, 2009).

45. Thomas Bender, "Introduction: Historians, the National, and the Plenitude of Narra-

tives," in *Rethinking American History in a Global Age*, ed. Thomas Bender (Berkeley: University of California, 2002), 8. Also see, for example, David Thelen, "The Nation and Beyond: Transnational Perspectives on United States History," *Journal of American History* 86 (1999): 965–75; Ian Tyrrell, *Transnational Nation: United States History in Global Perspective since 1789* (New York: Palgrave Macmillan, 2007); Ian Tyrrell, "American Exceptionalism in an Age of International History," *American Historical Review* 96 (1991): 1031–55; and Thomas Bender, *A Nation Among Nations: America's Place in World History* (New York: Hill and Wang, 2006).

46. Interview with William Berman by Frederick C. Durant III, December 10, 1984, 50, Comsat History Project, Series III: Comsat History Files, Comsat Corporation Collection, Milton S. Eisenhower Library Special Collections, Johns Hopkins University, Baltimore, Md.

47. David Halberstam, *The Best and the Brightest* (1972; New York: Modern Library, 2001).

48. See, for example, Bijker, Hughes, and Pinch, *Social Construction of Technological Systems*; Bijker and Law, *Shaping Technology/Building Society*; Leo Marx and Merritt Roe Smith, *Does Technology Drive History? The Dilemma of Technological Determinism* (Cambridge, Mass.: MIT Press, 1994); Douglas, *Inventing American Broadcasting*; Pamela E. Mack, *Viewing the Earth: The Social Construction of the Landsat Satellite System* (Cambridge, Mass.: MIT Press, 1990); Ronald R. Kline, *Consumers in the Country: Technology and Social Change in Rural America* (Baltimore, Md.: Johns Hopkins University Press, 2000); David Edgerton, *The Shock of the Old: Technology and Global History since 1900* (Oxford: Oxford University Press, 2007); David Nye, *Technology Matters: Questions to Live With* (Cambridge, Mass.: MIT Press, 2006); Ruth Cowan, *More Work for Mother: The Ironies of Household Technology from the Open Hearth to the Microwave* (New York: Basic Books, 1983); Michele Martin, *"Hello Central?" Gender, Technology, and Culture in the Formation of Telephone Systems* (Montreal: McGill-Queen's University Press, 1991); Claude S. Fischer, *America Calling: A Social History of the Telephone to 1940* (Berkeley: University of California Press, 1992); Thomas Misa, *A Nation of Steel: The Making of Modern America* (Baltimore, Md.: Johns Hopkins University Press, 1995); Hugh Richard Slotten, *Radio's Hidden Voice: The Origins of Public Broadcasting in the United States* (Urbana: University of Illinois Press, 2009); and Slotten, *Radio and Television Regulation*. On the importance of the social construction of technology approach to the history of communications, see Balbi and John, "Point-to-Point," 47–49.

49. John Fousek, *To Lead the Free World: American Nationalism and the Cultural Roots of the Cold War* (Chapel Hill: University of North Carolina Press, 2000). Also see Jenifer Van Vleck's similar analysis for American aviation in *Empire of the Air*, 10–11. And on US globalism in general, see Daniel Immerwahr, *How to Hide an Empire: A History of the Greater United States* (New York: Farrar, Straus and Giroux, 2019); and Ekbladh, *Great American Mission*.

50. Latham, *Modernization as Ideology*. Also see David C. Engerman, Nils Gilman, Michael E. Latham, and Mark Haefele, eds., *Staging Growth: Modernization, Development, and the Cold War* (Amherst: University of Massachusetts Press, 2003).

Chapter 1 · US Industry, the Cold War, and the Development of Satellite Communications

1. Arthur C. Clarke, "Extra-Terrestrial Relays: Can Rocket Stations Give World-Wide Radio Coverage?," reproduced in John M. Logsdon, ed., *Exploring the Unknown: Selected Documents in the History of the U.S. Civilian Space Program*, vol. 3, *Using Space* (Washington, D.C.: NASA, 1998), 12–15.

2. On the postwar activities of military agencies, see Walter A. McDougall, . . . *The Heavens and the Earth: A Political History of the Space Age* (Baltimore, Md.: Johns Hopkins University Press, 1997), 99–111; and David M. Hart, *Forged Consensus: Science, Technology, and Economic Policy in the United States, 1921–1953* (Princeton, N.J.: Princeton University Press, 1998).

3. Hart, *Forged Consensus*, 185.

4. Hart, *Forged Consensus*, 194–95.

5. McDougall, *Heavens and the Earth*, 104–11.

6. McDougall, *Heavens and the Earth*, 104–11.

7. McDougall, *Heavens and the Earth*, 118; Roger Launius, James Rodger Fleming, and David DeVorkin, eds. *Globalizing Polar Science: Reconsidering the International Polar and Geophysical Years* (New York: Palgrave Macmillan, 2010); Allan A. Needell, *Science, Cold War, and the American State: Lloyd V. Berkner and the Balance of Professional Ideals* (Amsterdam: Harwood Academic, 2000); Susan Barr and Cornelia Lüdecke, eds., *The History of the International Polar Years (IPYs)* (New York: Springer, 2010). On the IGY, also see Benjamin W. Goossen, "A Benchmark for the Environment: Big Science and 'Artificial' Geophysics in the Global 1950s," *Journal of Global History* 15 (2020): 149–68.

8. McDougall, *Heavens and the Earth*, 104–11.

9. McDougall, *Heavens and the Earth*, 118–23.

10. On the background to Sputnik in the United States, see especially R. Cargill Hall, "The Eisenhower Administration and the Cold War: Framing American Astronautics to Serve National Security," in *Organizing for the Use of Space: Historical Perspectives on a Persistent Issue*, ed. Roger D. Launius (San Diego, Calif: Univelt, 1995), 54–58; McDougall, *Heavens and the Earth*, 118–24; Dwayne A. Day, "A Strategy for Reconnaissance: Dwight D. Eisenhower and Freedom of Space," in *Eye in the Sky: The Story of the Corona Spy Satellites*, ed. Dwayne A. Day, John M. Logsdon, and Brian Lattell (Washington, D.C.: Smithsonian Institution Press, 1998), 119–42; and Roger D. Launius, John M. Logsdon, and Robert W. Smith, eds., *Reconsidering Sputnik: Forty Years since the Soviet Satellite* (London: Routledge, 2014). On military reconnaissance satellites, see Paul B. Stares, *The Militarization of Space: U.S. Policy, 1945–84* (Ithaca, N.Y.: Cornell University Press, 1985); and Kevin C. Ruffner, ed., *Corona: America's First Satellite Program* (Washington, D.C.: Center for the Study of Intelligence, Central Intelligence Agency, 1995).

11. David J. Whalen, "Billion Dollar Technology: A Short Historical Overview of the Origins of Communication Satellite Technology, 1945–1965," in *Beyond the Ionosphere: Fifty Years of Satellite Communication*, ed. Andrew J. Butrica (Washington, D.C.: NASA, 1997), 97.

12. "Additional Position Papers for the United States Delegation to the Extraordinary Administrative Radio Conference for Space Radio Communication and Radio Astronomy," August 23, 1963, 26, TEL 6-1, Space Communication Frequencies, 2/1/63, ITU folder, box 3655, Central Foreign Policy File, 1963, General Records of the Department of State, Record Group (RG) 59, National Archives and Records Administration (NARA), College Park, Md.

13. David K. van Keuren, "Moon in Their Eyes: Moon Communication Relay at the Naval Research Laboratory, 1951–1962," in Butrica, *Beyond the Ionosphere*, 9–18.

14. See introduction in Butrica, *Beyond the Ionosphere*, xix.

15. Headrick, *Invisible Weapon*, 201–2, quotation on 202.

16. Donald C. Elder, "Something of Value: Echo and the Beginnings of Satellite Communications," in Butrica, *Beyond the Ionosphere*, 34–35. On Pierce's cost estimates, see Whalen, "Billion Dollar Technology," 96. On the importance of science fiction and other imaginative literature in stimulating and legitimating the early development of the US space program, see Howard E. McCurdy, *Space and the American Imagination* (Washington, D.C.: Smithsonian Institution Press, 1997). For a discussion of AT&T's research relating to satellite communications, see David J. Whalen, *The Origins of Satellite Communications, 1945–1965* (Washington, D.C.: Smithsonian Institution Press, 2002), 32–40. For a discussion of relevant developments at Bell Laboratories, see F. R. Kappel to T. Keith Glennan, December 14, 1960, reprinted in Logsdon, *Exploring the Unknown*, 3:52–54.

17. On the Eisenhower administration's response to Sputnik, see Robert A. Divine, *The Sputnik Challenge* (New York: Oxford University Press, 1993); Rip Bulkeley, *The Sputniks Crisis and Early United States Space Policy: A Critique of the Historiography of Space* (Bloomington:

Indiana University Press, 1991); and James R. Killian Jr., *Sputnik, Scientists, and Eisenhower: A Memoir of the First Special Assistant to the President for Science and Technology* (Cambridge, Mass.: MIT Press, 1977). On the early history of NASA, see Enid Curtis Bok Schoettle, "The Establishment of NASA," in *Knowledge and Power: Essays on Science and Government*, ed. Sanford A. Lakoff (New York: Free Press, 1966), 162–270; Robert Rosholt, *An Administrative History of NASA, 1958–1963* (Washington, D.C.: NASA, 1966); and Allison Griffith, *The National Aeronautics and Space Act: A Study of the Development of Public Policy* (Washington, D.C.: Public Affairs Press, 1962). For a general discussion of the creation of NASA, changes in other agencies of the federal government, and the expansion of federal spending for high-technology weapons, see John M. Logsdon, *The Decision to Go to the Moon: Project Apollo and the National Interest* (Chicago: University of Chicago Press, 1970), 12–91; and McDougall, *Heavens and the Earth*, 151, 169, 176, 228.

18. David N. Spires and Rick W. Sturdevant, "From Advent to Milstar: The U.S. Air Force and the Challenges of Military Satellite Communications," in Butrica, *Beyond the Ionosphere*, 66–67; Whalen, *Origins of Satellite Communications*, 41–54; House Committee on Science and Astronautics, *Satellites for World Communication*, 86th Cong., 1st sess. (May 7, 1959), 2–4; James Schwoch, *Global TV: New Media and the Cold War, 1946–69* (Urbana: University of Illinois Press, 2009), 123–25.

19. For a description of the three original projects—Steer, Tackle, and Decree—and the decision to create Advent, see House Subcommittee on Military Operations of the Committee on Government Operations, *Military Communications Satellite Program: Hearings before the Subcommittee of the Committee on Government Operations*, 88th Cong., 1st sess. (April 23, 1963), 14.

20. Testimony of Admiral Jack Dorsey, House Subcommittee on Military Operations, *Military Communications Satellite Program*, 14–16, 23.

21. John H. Rubel to George Miller, July 10, 1962, Communications Satellite Program (DOD) folder, box 7, Directors Comsat Records 1962–66, Records of the Office of Emergency Preparedness, RG 396, NARA.

22. Whalen, *Origins of Satellite Communications*, 47–69. Also see Joan Lisa Bromberg, *NASA and the Space Industry* (Baltimore, Md.: Johns Hopkins University Press, 1999), 47–49.

23. For a discussion of AT&T's research relating to satellite communications during this period, see Whalen, *Origins of Satellite Communications*, 44–69. For a discussion of Bell Laboratories, see Kappel to Glennan, December 14, 1960.

24. Elder, "Something of Value," 35–38; Craig B. Waff, "Project Echo, Goldstone, and Holmdel: Satellite Communications as Viewed from the Ground Station," in Butrica, *Beyond the Ionosphere*, 43.

25. Whalen, *Origins of Satellite Communication*, 44–52, 56–57. Also see David J. Whalen, *The Rise and Fall of Comsat: Technology, Business, and Government in Satellite Communications* (Basingstoke: Palgrave Macmillan, 2014), chap. 1.

26. John R. Pierce, "Exotic Radio Communications," September 1959, reprinted in Logsdon, *Exploring the Unknown*, 3:30.

27. Whalen, *Origins of Satellite Communications*, 57–66.

28. Whalen, *Origins of Satellite Communications*, 48, 64–69.

29. For a discussion of AT&T's choice of system, see J. E. Dingman to Ben F. Waple, March 21, 1961, reprinted in Logsdon, *Exploring the Unknown*, 3:57; and Whalen, *Origins of Satellite Communications*. For a complete comparison of the two different systems from a slightly later period (1963), see "Annex 2 to letter [from Ad Hoc Committee of the Telecommunications Commission of the CEPT] to be addressed to the Satellite Communications Corporation," June 25, 1963, Telecommunications: TEL 3, Organizations and Conferences, CEPT folder, box 3652, Central Foreign Policy File, 1963, General Records of the Department of State, RG 59, NARA.

30. On Hughes's activities, see H. A. Rosen and D. D. Williams, "Commercial Communica-

tions Satellite," January 1960, reprinted in Logsdon, *Exploring the Unknown*, 3:35–39; and Whalen, *Origins of Satellite Communications*, 61.

31. "Policy Questions Concerning Space Communications Systems (FCC Inter-office Memorandum)," December 5, 1960, microfilm reel 1, Records of the Federal Communications Commission (FCC), John F. Kennedy (JFK) Presidential Library, Boston, Mass., 7, 15.

32. "Policy Questions Concerning Space Communication Systems," 3 (quotation from Communications Act), 7, 15.

33. J. D. Hunley, ed., *The Birth of NASA: The Diary of T. Keith Glennan* (Washington, D.C.: NASA, 1993), 190.

34. "Major Actions of the President's Science Advisory Committee: November 1957–January 1961," January 13, 1961, Office of Science and Technology, General 1961–62, folder, box 284–85 (combined), National Security Files, Departments and Agencies, Papers of President Kennedy, JFK Presidential Library, Boston, Mass.

35. Hunley, *Birth of NASA*, 194, 260, 292. For a description of Glennan and his policies, see especially Roger D. Launius, "Early U.S. Civil Space Policy, NASA, and the Aspiration of Space Exploration," in *Organizing for the Use of Space*, 82–85. On Eisenhower's general policies regarding the space program, see David Callahan and Fred I. Greenstein, "The Reluctant Racer: Eisenhower and U.S. Space Policy," in *Spaceflight and the Myth of Presidential Leadership*, ed. Roger D. Launius and Howard E. McCurdy (Urbana: University of Illinois Press, 1997), 15–50; and Vernon van Dyke, *Pride and Power: The Rationale of the Space Program* (Urbana: University of Illinois Press, 1964), 22.

36. "Memorandum for Conference on Communications Satellite Development," December 7, 1960, reprinted in Logsdon, *Exploring the Unknown*, 3:40.

37. White House Press Secretary, "Statement by the President," December 30, 1960, reprinted in Logsdon, *Exploring the Unknown*, 3:42.

38. Attorney General Rogers's comments recorded by Robert G. Nunn, memorandum for the record, December 23, 1960, folder 013925, NASA History Office, Washington, D.C. For a discussion of the "open door" metaphor in communications policy, see Andrew L. Russell, *Open Standards and the Digital Age: History, Ideology, and Networks* (Cambridge: Cambridge University Press, 2014).

39. Hunley, *Birth of NASA*, 285–86, 290, 302–3, 306, quotation on 286.

40. Hunley, *Birth of NASA*, 285–86, 290, 301–3, 306, quotations on 301.

41. Thomas E. Will, *Telecommunications Structure and Management in the Executive Branch of Government, 1900–1970* (Boulder, Colo.: Westview Press, 1978), 9, 13–16. On the establishment of IRAC, also see Louise M. Benjamin, "Regulating the Government Airwaves: Creation of the Interdepartmental Radio Advisory Committee (IRAC)," *Journal of Broadcasting and Electronic Media* 51 (September 2007): 498–515.

42. Will, *Telecommunications Structure and Management*, 18–19.

43. Will, *Telecommunications Structure and Management*, 18–19.

44. "TCC Working Group: Space Satellite Communications—Summary Record of Meeting," October 17, 1960, microfilm roll 1, Records of the FCC, JFK Presidential Library, Boston, Mass.

45. "TCC Working Group: Space Satellite Communications." For questions raised by TCC, see "Some Problems Involved in Formulation of United States Government Policy on Communications by Space Satellite Relays," September 26, 1960, microfilm roll 1, Records of the FCC, JFK Presidential Library, Boston, Mass.

46. Senate Committee on Aeronautical and Space Sciences, *Policy Planning for Space Telecommunications*, 86th Cong., 2d sess. (November 9, 1960), 87, 99–100, 120.

47. Senate Committee on Aeronautical and Space Sciences, *Policy Planning for Space Telecommunications*, 100.

48. Senate Committee on Aeronautical and Space Sciences, *Policy Planning for Space Telecommunications*, 33.

49. Nandasiri Jasentuliyana, "Regulatory Functions of ITU in the Field of Space Telecommunications," *Journal of Air Law and Commerce* 34 (1968): 62–78.

50. J. Henry Glazer, "The Law-Making Treaties of the International Telecommunication Union through Time and in Space," *Michigan Law Review* 60 (January 1962): 285–86n59.

51. James P. Gleason (NASA) to Lyndon B. Johnson, July 12, 1960, reprinted in Senate Committee on Aeronautical and Space Sciences, *Policy Planning for Space Telecommunications*, 128.

52. In 1960, Andrew Haley, a lawyer representing the American Rocket Society, with a background as an amateur rocketeer, recalled how officials had mostly ignored his repeated pleas for space frequencies during 1956. "Then a year later," according to Haley, "the Russians put out a Sputnik and everybody change[d] their opinion all of a sudden." Haley testimony in *Official Report of Proceedings before the Federal Communications Commission: In the Matter of American Telephone and Telegraph Co., . . .*, July 18, 1960, 5230, box 6007, docket 11866, Docketed Case Files, Records of the FCC, RG 173, NARA. On Haley, see McDougall, *Heavens and the Earth*, 188.

53. George Arthur Codding Jr., *The International Telecommunication Union: An Experiment in International Cooperation* (Leiden, Netherlands: Brill, 1952), 13–179; Gabriele Balbi and Andreas Fickers, eds., *History of the International Telecommunication Union* (Berlin: De Gruyter Oldenbourg, 2020). On early radio conferences before World War II, see especially James Schwoch, *The American Radio Industry and Its Latin American Activities, 1900–1939* (Urbana: University of Illinois Press, 1990), 56–95; and James Schwoch, "The American Radio Industry and International Communications Conferences, 1919–1927," *Historical Journal of Film, Radio, and Television* 7 (October 1987): 289–309.

54. Codding, *International Telecommunication Union*, 13–179.

55. James G. Savage, *The Politics of International Telecommunications Regulation* (Boulder, Colo.: Westview Press, 1989), 38–40; George A. Codding Jr. and Anthony M. Rutkowski, *The International Telecommunication Union in a Changing World* (Dedham, Mass.: Artech House, 1982), 21–24; Christiane Berth, "ITU, the Development Debate, and Technical Cooperation in the Global South, 1950–1992," in Balbi and Fickers, *History of the International Telecommunication Union*, 69–95, on 71. The best general historical discussion of the 1947 conference is Schwoch, *Global TV*, 17–30. For details about how Switzerland "had to readjust its relationship with the ITU," see Anne-Katrin Weber, Roxane Gray, and Marie Sandoz, "ITU Exhibitions in Switzerland: Displaying the 'Big Family of Telecommunications,' 1960s-1970s," in Balbi and Fickers, *History of the International Telecommunication Union*, 237–64.

56. Savage, *Politics of International Telecommunications Regulation*, 38–40; Codding and Rutkowski, *International Telecommunication Union in a Changing World*, 21–24.

57. Savage, *Politics of International Telecommunications Regulation*, 40 41, 70–74; Codding and Rutkowski, *International Telecommunication Union in a Changing World*, 25–26.

58. Savage, *Politics of International Telecommunications Regulation*, 41, 75–77; Codding and Rutkowski, *International Telecommunication Union in a Changing World*, 26–31.

59. Harold K. Jacobson, "ITU: A Potpourri of Bureaucrats and Industrialists," in *The Anatomy of Influence: Decision Making in International Organization* (New Haven, Conn.: Yale University Press, 1973), 97.

60. Savage, *Politics of International Telecommunications Regulation*, 75–77; Codding and Rutkowski, *International Telecommunication Union in a Changing World*, 31–39, 212.

61. Robert G. Weston, "The United Nations in the World Outlook of the Soviet Union and the United States," in *Soviet and American Policies in the United Nations: A Twenty-Five Year Perspective*, ed. Alvin Z. Rubinstein and George Ginsburgs (New York: New York University Press, 1971), 6–8, 18, quotations on 8 and 18.

62. Paul Gordon Lauren, "The Diplomats and Diplomacy of the United Nations," in *The Diplomats, 1939–1979*, ed. Gordon A. Craig and Francis L. Loewenheim (Princeton, N.J.: Princeton University Press, 1994), 469.

63. Senate Committee on Aeronautical and Space Sciences, *Communication Satellites: Technical, Economic, and International Developments*, staff report prepared by Donald R. Mac-Quivey, 87th Cong., 2d sess. (1962), Committee Print (CIS-NO:S0525), 18–19.

64. During most of this period, the State Department was organized into regional bureaus and several key specialized bureaus headed by assistant secretaries and deputy assistant secretaries. Besides the Bureau of Economic Affairs, these specialized bureaus included the Bureau of International Organization Affairs, the Bureau of Congressional Relations, and the Bureau of Public Affairs. Besides the secretary, undersecretary, assistant secretaries, and deputy assistant secretaries, the other major officers included the second undersecretary (either undersecretary for political affairs or undersecretary for economic affairs), the deputy undersecretary for political affairs, and the deputy undersecretary for administration. The deputy undersecretary for political affairs had an important role of working with the Defense Department and the intelligence agencies. See especially memorandum from Secretary of State Rusk to President Johnson, December 31, 1964, S/S-Ball Files, Lot 74 D 272, Under Secretary Ball—1964, General Records of the Department of State, RG 59, NARA, reprinted as document 16 in US State Department, *Foreign Relations of the United States, 1964–68*, vol. 33, *Organization and Management of Foreign Policy, United Nations, the Department of State and the Coordination and Supervision of U.S. Foreign Policy* (Washington, D.C.: Government Printing Office, 2004).

65. Senate Committee on Aeronautical and Space Sciences, *Radio Frequency Control in Space Telecommunications*, report prepared by Edward Wenk, 86th Cong., 2d sess. (1960), Committee Print (CIS-NO: S2131), 23.

66. House Subcommittee of the Committee on Interstate and Foreign Commerce, *Spectrum Allocation: Hearings on Allocation of Radio Spectrum between Federal and Non-Federal Government Users*, 86th Cong., 1st sess. (June 8, 1959), 105–20. On the establishment of IRAC, see Benjamin, "Regulating the Government Airwaves," 498–515.

67. Executive Order 10460 quoted in Senate Communications Subcommittee of the Committee on Commerce, *Space Communications and Allocation of Radio Spectrum: Hearings on Space Communications and S.J. Res. 32*, 87th Cong., 1st sess. (August 23, 1961), 113.

68. House Subcommittee of the Committee on Interstate and Foreign Commerce, *Spectrum Allocation*, 119–20; Herbert I. Schiller, "The Increasing Military Influence in the Governmental Sector of Communications in the United States," *Administrative Law Review* 303 (1966–67): 303–18. The Office of Civil and Defense Mobilization was created as a response to Sputnik by merging the Office of Defense Mobilization and the Federal Civil Defense Agency. See Will, *Telecommunications Structure and Management*, 26–27, 37–39.

69. House Subcommittee of the Committee on Interstate and Foreign Commerce, *Spectrum Allocation*, 119–20. Also see Schiller, "Increasing Military Influence," 303–18.

70. Will, *Telecommunications Structure and Management*, 31, 34, 37–39, 58; "electromagnetic war" quotation on 58, second CIA quotation on 31.

71. H. Leslie Hoffman (member of the Spectrum Study Committee, Electronic Industries Association) testimony in House Subcommittee of the Committee on Interstate and Foreign Commerce, *Spectrum Allocation*, 71.

72. Untitled report, February 10, 1961, attached to letter from John O. Pastore to John Fitzgerald Kennedy, March 6, 1961, 2, ND 3 Communications—Electronics folder, box 4, National Security—Defense ND 3 Collection, Lyndon B. Johnson Presidential Library, Austin, Texas.

73. Pastore to Kennedy, March 6, 1961. On some of the major studies undertaken during this period that focused on the possibility of centralizing communications policy, see especially discussion in "Statement by Senator Vance Hartke in support of Senate Joint Resolution 32," in

Senate Communications Subcommittee, *Space Communications and Allocation of Radio Spectrum*, 99. Also see "Federal Administration Problems of National Telecommunications," box 6005, docket 11866, Docketed Case Files, Records of the FCC, RG 173, NARA.

74. Senate Communications Subcommittee, *Space Communications and Allocation of Radio Spectrum*, 147.

75. Senate Communications Subcommittee, *Space Communications and Allocation of Radio Spectrum*, 149.

76. Donald Beelar quoted in Senate Communications Subcommittee, *Space Communications and Allocation of Radio Spectrum*, 168.

77. Senate Communications Subcommittee, *Space Communications and Allocation of Radio Spectrum*, 159

78. Senate Communications Subcommittee, *Space Communications and Allocation of Radio Spectrum*, 161.

79. Untitled report, February 10, 1961, 2.

80. Senate Communications Subcommittee, *Space Communications and Allocation of Radio Spectrum*, 161.

81. Senate Communications Subcommittee, *Space Communications and Allocation of Radio Spectrum*, 168.

82. Senate Communications Subcommittee, *Space Communications and Allocation of Radio Spectrum*, 168. Critics pressured the Eisenhower administration to respond aggressively to Nikita Khrushchev's threats during the period of increased tension following Sputnik. See Kenneth Alan Osgood, *Total Cold War: Eisenhower's Secret Propaganda Battle Home and Abroad* (Lawrence: University of Kansas, 2008), 323–53; and Divine, *Sputnik Challenge*.

83. John C. Doerfer testimony in House Subcommittee of the Committee on Interstate and Foreign Commerce, *Spectrum Allocation*, 65. The role of the radio spectrum in the Cold War was particularly important during the period of increased tension following Sputnik, when critics pressured the Eisenhower administration to respond aggressively to Khrushchev's threats. See Osgood, *Total Cold War*, 323–53; and Divine, *Sputnik Challenge*. The growing demand for consumer electronics after World War II also conflicted with the dramatic increase in military demands for radio frequencies during the Cold War. Key members of Congress pressured the military to release frequencies for private use. Members of the Senate Interstate and Foreign Commerce Committee were particularly concerned about finding ways to support the full potential of television broadcasting during the 1950s, especially by opening up bands of frequencies for the new technology. But the military services resisted. The FCC had first assigned television broadcasting to the VHF (very high frequency) spectrum. But the limited band of frequencies assigned to television could not provide high-quality, multiple services to all regions of the country. During the early 1950s, the commission attempted to move television to higher channels made available after the war (in the UHF band). New UHF stations, however, could not compete with established VHF broadcasters. The FCC chose not to force all VHF broadcasters to move to the UHF band. VHF television did not have room to expand in the radio spectrum mainly because the military services controlled adjacent frequencies. IRAC supported the military when it rejected requests made by Congress and the FCC to trade government/military VHF frequencies for some of the UHF channels originally assigned to television. The military argued that it would cost approximately five billion dollars to move its operations to the UHF band; the change would also cause a dangerous disruption to defense operations. See Will, *Telecommunications Structure and Management*, 26–27, 37–39.

84. Despite the willingness of Soviets during the late 1950s to participate in the UN and its specialized agencies, the United States still dominated in terms of personnel. See *Report of the Chairman of the United States Delegation to the Administrative Radio Conference of the International Telecommunication Union, Geneva, Switzerland—August 17, 1959 through December 21,*

1959, annex A, 399.20-ITU/5–260 folder, box 837, Central Decimal File, 1960–63, General Records of the Department of State, RG 59, NARA.

85. See *Report of the Chairman of the United States Delegation to the Administrative Radio Conference.*

86. Jasentuliyana, "Regulatory Functions of ITU," 66–67. The Soviets also developed medium-altitude (not geostationary) satellites with special orbits more appropriate for a large country located in the far north. The communications satellites the Soviets developed during the mid-1960s, the Molniya series, had a highly elliptical orbit oriented in such a way that when they were nearest to the Earth, they would appear to be relatively stationary over Soviet territory for long periods.

87. Senate Committee on Aeronautical and Space Sciences, *Radio Frequency Control in Space Telecommunications*, 50–51. On the development by the United States military of tracking and Earth stations, see David Christopher Arnold, *Spying from Space: Constructing America's Satellite Command and Control Systems* (College Station: Texas A&M University Press, 2005).

88. Action memorandum from the director of the Office of International Scientific Affairs (Rollefson) to Secretary of State Rusk, July 11, 1963, Central Files 1960–63, SP 10 US, General Records of the Department of State, RG 59, NARA, reprinted as document 382 in US State Department, *Foreign Relations of the United States, 1961–63*, vol. 25, *Organization of Foreign Policy, Information Policy, United Nations, Scientific Matters, U.S. Space Program.*

89. Senate Committee on Aeronautical and Space Sciences, *Radio Frequency Control in Space Telecommunications*, 50–51.

90. Senate Communications Subcommittee, *Space Communications and Allocation of Radio Spectrum*, 151.

91. Senate Committee on Aeronautical and Space Sciences, *Policy Planning for Space Telecommunications*, 51.

92. Frederick W. Ford (FCC chairman) to Lyndon B. Johnson, June 29, 1960, reproduced in Senate Committee on Aeronautical and Space Sciences, *Policy Planning for Space Telecommunications*, 133. For the specific frequencies, see *Radio Regulations: Additional Radio Regulations, Additional Protocol, Resolutions and Recommendations, Geneva, 1959* (Geneva: General Secretariat of the International Telecommunication Union, 1959), http://search.itu.int/history/History DigitalCollectionDocLibrary/4.85.43.en.100.pdf, 50, 53, 57, 65, 70, 72, 80, 82, 89, 91, 95, 97.

93. W. B. Franke (Dept. of Navy) to Lyndon B. Johnson, June 13, 1960, reproduced in Senate Committee on Aeronautical and Space Sciences, *Policy Planning for Space Telecommunications*, 161.

94. *Report of the Chairman of the United States Delegation to the Administrative Radio Conference.*

95. Vladislav M. Zubok, *A Failed Empire: The Soviet Union in the Cold War from Stalin to Gorbachev* (Chapel Hill: University of North Carolina Press, 2007), 94; William Taubman, *Khrushchev: The Man and His Era* (New York: Norton, 2003), 347–48; Weston, "United Nations in the World Outlook," 12–13; Vladislav Zubok and Constantine Pleshakov, *Inside the Kremlin's Cold War: From Stalin to Khrushchev* (Cambridge Mass.: Harvard University Press, 1996), 174–85; Aleksandr Fursenko and Timothy Naftali, *Khrushchev's Cold War: The Inside Story of an American Adversary* (New York: Norton, 2006), 22–32, 241; Alexander Dallin, *The Soviet Union at the United Nations: An Inquiry into Soviet Motives and Objectives* (London: Methuen, 1962), 39–40, 115–22.

96. Dallin, *Soviet Union at the United Nations*, 130.

97. Weston, "United Nations in the World Outlook," 13

98. Weston, "United Nations in the World Outlook," 13.

99. Harold Jacobson, "Decolonization," in Rubinstein and Ginsburgs, *Soviet and American Policies in the United Nations*, 79–82; Dallin, *Soviet Union at the United Nations*, 40.

100. Report quoted in Senate Committee on Aeronautical and Space Sciences, *Policy Planning for Space Telecommunications*, 98. Also see Berth, "ITU, the Development Debate," 72–76.

101. Berth, "ITU, the Development Debate," 75.

102. Senate Committee on Aeronautical and Space Sciences, *Policy Planning for Space Telecommunications*, 98.

103. Osgood, *Total Cold War*, especially 323–53, Eisenhower quoted on 347; Chester J. Pach Jr., "Introduction: Thinking Globally and Acting Locally," in *The Eisenhower Administration, the Third World, and the Globalization of the Cold War*, ed. Kathryn C. Statler and Andrew L. Johns (New York: Rowman and Littlefield, 2006); Zubok, *Failed Empire*, 139.

104. The idea that the United States should take the lead in developing a global satellite communications system serving Cold War aims, especially by linking non-Western or non-aligned countries to the United States, was first expressed in the staff report written in fall 1960, at the request of the Senate Committee on Aeronautical and Space Sciences, chaired by Senator Lyndon Johnson. See Senate Committee on Aeronautical and Space Sciences, *Policy Planning for Space Telecommunications*, 87, 99–100, 120.

Chapter 2 · The Kennedy Administration and the Communications Satellite Act of 1962

1. Hugh Sidey, *John F. Kennedy, President* (New York: Atheneum, 1964), 98. Also see John M. Logsdon, *John F. Kennedy and the Race to the Moon* (New York: Palgrave Macmillan, 2010), 1–13.

2. Christopher A. Preble, *John F. Kennedy and the Missile Gap* (DeKalb: Northern Illinois University Press, 2004).

3. Logsdon, *John F. Kennedy and the Race to the Moon*, 37.

4. Kennedy's Inaugural Address, January 20, 1961, "Historic Speeches," John F. Kennedy (JFK) Presidential Library and Museum, jfklibrary.org/learn/about-jfk/historic-speeches/inaugural-address.

5. On the Laos crisis, see Laos: General, 1961: January–March, digital identifier JFKPOF-121-007, JFK Presidential Library, https://www.jfklibrary.org/asset-viewer/archives/JFKPOF/121/JFKPOF-121-007. And see Logsdon, *John F. Kennedy and the Race to the Moon*, 39.

6. For digital copy of the Kennedy's 1961 State of the Union Address, January 30, 1961, see digital identifier JFKPOF-034-003, JFK Presidential Library, https://www.jfklibrary.org/asset-viewer/archives/JFKPOF/034/JFKPOF-034-003.

7. William Taubman, *Khrushchev: The Man and His Era* (London: Free Press, 2005), 487; A. A. Fursenko and Timothy J. Naftali, *Khrushchev's Cold War: The Inside Story of an American Adversary* (New York: Norton, 2006), 342. On the space proposal in the inaugural address, see comments with a draft of the proposal made by the administration on the topic in April with Document I-36 in John M. Logsdon, "The Development of International Space Cooperation," in *Exploring the Unknown: Selected Documents in the History of the U.S. Civilian Space Program*, vol. 2, *External Relationships*, ed. John M. Logsdon and Dwayne A. Day (Washington, D.C.: NASA, 1996), 143–47.

8. Gerard J. DeGroot, *Dark Side of the Moon: The Magnificent Madness of the American Lunar Quest* (New York: New York University Press, 2006), 135.

9. David Tal, *The American Nuclear Disarmament Dilemma, 1945–1963* (Syracuse, N.Y.: Syracuse University Press, 2008), 180–81; Taubman, *Khrushchev*, 502–3. For quotation, see Fursenko and Naftali, *Khrushchev's Cold War*, 379.

10. National Security Council, "Statement of Preliminary U.S. Policy on Outer Space" (NSC 5814/1), August 18, 1958, reprinted as document 442 in US State Department, *Foreign Relations of the United States, 1958–1960*, vol. 2, *United Nations and General International Matters, Outer Space* (Washington, D.C.: Government Printing Office), 851.

11. In a letter to Soviet official Nikolai Bulganin on February 15, 1958, Eisenhower proposed

"wholly eliminating the newest types of weapons which use outer space for human destruction." Eisenhower quoted in National Security Council, "Statement of Preliminary U.S. Policy on Outer Space," 852.

12. National Security Council, "Statement of Preliminary U.S. Policy on Outer Space," 853.

13. Department of State to UN Mission, August 18, 1958, reprinted as document 443 in US State Department, *Foreign Relations of the United States, 1958–1960*, vol. 2.

14. UN Mission to Department of State, November 18, 1958, reprinted as document 449 in US State Department, *Foreign Relations of the United States, 1958–1960*, vol. 2.

15. Department of State to UN Mission, November 11, 1958, reprinted in US State Department, *Foreign Relations of the United States, 1958–1960*, 2:867.

16. Department of State to UN Mission, November 19, 1958, reprinted as document 450 in US State Department, *Foreign Relations of the United States, 1958–1960*, vol. 2; "Position Paper Prepared for the Fourteenth Regular Session of the United Nations General Assembly," September 9, 1959, reprinted as document 460 in US State Department, *Foreign Relations of the United States, 1958–1960*, 2:888. Also see Matthew J. Von Bencke, *The Politics of Space: A History of U.S.-Soviet/Russian Competition and Cooperation in Space* (Boulder, Colo.: Westview Press, 1997), 42–43.

17. Department of State to "Certain Diplomatic Missions," April 28, 1959, reprinted as document 458 in US State Department, *Foreign Relations of the United States, 1958–1960*, 2:885.

18. Dean Rusk, memorandum for the president, n.d. (early 1961), microfilm reel 16, US Office of Science and Technology Records, JFK Presidential Library, Boston, Mass.

19. Rusk memorandum (early 1961).

20. Taubman, *Khrushchev*, 448–49; Vladislav Martinovich Zubok, *A Failed Empire: The Soviet Union in the Cold War from Stalin to Gorbachev* (Chapel Hill: University of North Carolina Press, 2006), 135; Fursenko and Naftali, *Khrushchev's Cold War*, 241–55.

21. Taubman, *Khrushchev*, 447.

22. Memorandum from Secretary of State Rusk to President Kennedy, February 2, 1961, reprinted as document 414 in *Foreign Relations of the United States, 1961–63*, vol. 25, *Organization of Foreign Policy, Information Policy, United States, Scientific Matters*; Logsdon, *John F. Kennedy and the Race to the Moon*, 161.

23. "Terms of Reference for Task Force on Possibilities for International Cooperation in Outer Space," February 9, 1961, attached to "Minutes—Meeting of Task Force on International Cooperation in Space Activities," February 27, 1961, Records of Government Agencies, Office of Science and Technology folder, reel 16, box 2 (microfilm open rolls), Records of the United States Office of Science and Technology, JFK Presidential Library, Boston, Mass.

24. Draft of report quoted by Logsdon, *John F. Kennedy and the Race to the Moon*, 163.

25. Logsdon, *John F. Kennedy and the Race to the Moon*, 163.

26. "Highlights of National Aeronautics and Space Council History," n.d., Administrative History of the Space Council (1961–1969) folder, box 23, NASC, Gen. Corresp., 1961–69, Records of Temporary Committees, Commission, and Boards, Record Group (RG) 220, National Archives and Records Administration (NARA), College Park, Md.

27. Logsdon, *John F. Kennedy and the Race to the Moon*, 29–30, quotation on 30; Robert A. Divine, *Sputnik Challenge* (New York: Oxford University Press, 1993).

28. Logsdon, *John F. Kennedy and the Race to the Moon*, 22, 29; Logsdon quotes Divine, *Sputnik Challenge*, 148–49.

29. Interview with Edward C. Welsh by Thomas Maxwell Safely, July 19, 1984, 32, Comsat History Project, Series III: Comsat History Files, Comsat Corporation Collection, Milton S. Eisenhower Library Special Collections, Johns Hopkins University, Baltimore, Md.

30. Logsdon, *John F. Kennedy and the Race to the Moon*, 18–22, 29–30.

31. John F. Kennedy to Overton Brooks, March 23, 1961, OS1 Aeronautical and Space Re-

search, 1961, folder, box 652, White House Central Subject File, Papers of President John F. Kennedy, JFK Presidential Library, Boston, Mass.

32. Welsh interview, 32.

33. Logsdon, *John F. Kennedy and the Race to the Moon*, 59.

34. Interview with John Johnson by Nina Gilden Seavey, March 13, 1986, 21, Comsat History Project, Comsat History Project, Series III: Comsat History Files, Comsat Corporation Collection, Milton S. Eisenhower Library Special Collections, Johns Hopkins University, Baltimore, Md.

35. Welsh interview, 47.

36. On Webb's involvement with the aeronautics industry, see James E. Webb Oral History, Interview 1, by T. H. Baker, April 29, 1969, p. 6, Lyndon B. Johnson (LBJ) Presidential Library, Austin, Texas.

37. W. Henry Lambright, *Powering Apollo: James E. Webb of NASA* (Baltimore, Md.: Johns Hopkins University Press, 1998), 5, 69–71, quotations on 69 and 71. For Webb quotation about the deal between Kerr and Johnson, see James E. Webb interview by Martin Collins, October 15, 1985, Smithsonian Institution National Air and Space Museum Oral History Project, Webb 8, tape 1, side 1, p. 18.

38. Lambright, *Powering Apollo*, 19, 77–83, 103, quotation on 103.

39. For the Wiesner Report, see "Report to the President-Elect of the Ad Hoc Committee on Space" (Committee 5 of the PSAC), January 12, 1961, folder record no. 006815, NASA History Office, Washington, D.C. For the Space Science Board report, see "Man's Role in the National Space Program," n.d. (released to government in March 1961), folder record no. 012083, NASA History Office, Washington, D.C. Also see Walter A. McDougall, . . . *The Heavens and the Earth: A Political History of the Space Age* (Baltimore, Md.: Johns Hopkins University Press, 1997), 309–10, 315.

40. For the Gardner Report, see "Report of the Air Force Space Study Committee," March 20, 1961, folder record no. 000755, NASA History Office, Washington, D.C. Overton Brooks to John F. Kennedy, March 9, 1961, OS1 Aeronautical and Space Research, 1961, folder, box 652, White House Central Subject File, Papers of President Kennedy, JFK Presidential Library, Boston, Mass.

41. McDougall, *Heavens and the Earth*, 315.

42. "Memorandum of Understanding between FCC and NASA on Respective Civil Space Communications Activities," February 28, 1961, Records of the Federal Communications Commission (FCC), roll 1, JFK Presidential Library, Boston, Mass. Also see text of same memorandum dated February 27, 1961, in DOD-NASA Agreements folder, box 10, NASC, Gen. Corresp., 1961–69, Records of Temporary Committees, Commission, and Boards, RG 220, NARA.

43. Interview with Asher Ende by Nina Gilden Seavey, August 28, 1985, 3–4, Comsat History Project, Series III: Comsat History Files, Comsat Corporation Collection, Milton S. Eisenhower Library Special Collections, Johns Hopkins University, Baltimore, Md.

44. Welsh interview, 1–2.

45. Memorandum for the vice president, Edward C. Welsh to Vice President Johnson, March 23, 1961, Budget—Space Agencies folder, box 13, NASC, Gen. Corresp., 1961–69, Records of Temporary Committees, Commission, and Boards, RG 220, NARA.

46. Welsh interview, 33.

47. Welsh interview, 1–2.

48. House Committee on Interstate and Foreign Commerce, *Communications Satellites, Part 2: Hearings on H.R. 10115 and H.R. 10138*, 87th Cong., 2nd sess. (1962), 618 (Webb quotation). On Webb's visionary effort to use the space program to bring the nation to a new frontier, see Lambright, *Powering Apollo*. On Webb's antimonopolist sentiments, see James E. Webb to R. R. Kappel, April 8, 1961, reprinted in John M. Logsdon, ed., *Exploring the Unknown: Selected*

Documents in the History of the U.S. Civilian Space Program, vol. 3, *Using Space* (Washington, D.C.: NASA, 1998), 59.

49. Memorandum, Franklyn W. Phillips (assistant to the NASA administrator) to Edward C. Welsh, April 11, 1961, Nat. Aeronautics and Space Administration folder, box 22, NASA, Gen. Corresp., 1961–69, Records of Temporary Committees, Commission, and Boards, RG 220, NARA.

50. Webb quoted in House Committee on Interstate and Foreign Commerce, *Communications Satellites, Part 1: Hearings on H.R. 108*, 87th Cong., 1st sess. (May 8–10 and July 13, 1961), 125, 127.

51. Welsh interview, 3.

52. Senate Committee on Aeronautical and Space Sciences, *Communication Satellites: Technical, Economic, and International Developments*, staff report prepared by Donald R. MacQuivey, 87th Cong., 2d sess. (1962), Committee Print (CIS-NO: S0525), 44. For Welsh's concerns about AT&T's monopoly control of international communications, see E. C. Welsh to the vice president, memorandum, March 13, 1963, Communication: Satellite folder, box 192, Vice President Collection, LBJ Presidential Library, Austin, Texas. Kennedy's science advisor, Jerome Wiesner, was particularly interested in communications policy and favored a strong government role to ensure that communications satellites supported international cooperation. For Wiesner's views about the role of the government in the development of communications satellites, see his report to President-Elect Kennedy, reprinted in House Committee on Science and Astronautics, *Defense Space Interests: Hearings before the Committee on Science and Astronautics*, 87th Cong., 1st sess. (1961), 22.

53. "Telesat: Telecommunication Satellite Business Planning Study," part one, October 1960, 12, no folder, carton 1, Beardsley Graham Papers, Bancroft Library, University of California, Berkeley.

54. Abe Silverstein to assistant directors et al., "Fiscal Year 1963 Preliminary Budget Estimates; Additional Information Concerning," March 1, 1961, NASA History Office, Washington, D.C.

55. "Telesat," part one, 4. This report also pointed out, "The Rand Corporation and Stanford Research Institute are currently conducting related studies for the National Aeronautics and Space Administration and others" (4).

56. Booz, Allen & Hamilton to Beardsley Graham, October 31, 1960, no folder, carton one, Beardsley Graham Papers, Bancroft Library, University of California, Berkeley. On Doerfer, see obituary in *New York Times*: "John Charles Doerfer, 87, Is Dead; Headed F.C.C. in Era of Scandals," June 8, 1992.

57. "Telesat," part one, 8, 19, 26, 36, 79–80 (quotations on 79–80).

58. "Telesat," part one, 4, 84–86, quotation on 86.

59. W. Wallace Kirkpatrick to Harold David Cohen, February 10, 1961, box 818, docket 13522, docket files, Records of the FCC, RG 173, NARA.

60. On reception of *Relay 1* broadcasts in Japan, see "Milestones: First Transpacific Reception of a Television (TV) Signal via Satellite, 1963," IEEE Milestones, IEEE Global History Network, *Engineering and Technology History Wiki*, November 23, 2009, http://www.ieeeghn.org/wiki/index.php/Milestones:First_Transpacific_Reception_of_a_Television_(TV)_Signal_via_Satellite,_1963.

61. Schwoch, *Global TV*, 128; "Telstar," The Tornadoes, The Hot 100, Billboard, December 22, 1962, https://www.billboard.com/charts/Hot-100/1962-12-22.

62. Memorandum, "Pornosat Project," June 1, 1961, Educational Satellite folder, carton 22, Beardsley Graham Papers, Bancroft Library, University of California-Berkeley.

63. "Initial World Reaction to Soviet 'Man in Space,'" April 21, 1961, Office of Research and Analysis, US Information Agency, Space Activities General 4/61–6/61 folder, box 307, Subjects, National Security Files, JFK Presidential Library, Boston, Mass.

64. Wernher von Braun to Vice President Johnson, April 29, 1961, Rec. No. 2682, LEK 8/4/7, NASA History Office, Washington, D.C.

65. "Presentation by the Administrator of the National Aeronautics and Space Administration [Webb] to President Kennedy," March 21, 1961, box 282, National Security Files, Departments and Agencies Series, NASA, 1961, JFK Presidential Library, reprinted as document 359 in US State Department, *Foreign Relations of the United States, 1961–63*, vol. 25.

66. For a good overview, see especially John M. Logsdon, "The Evolution of U.S. Space Policy and Plans," in *Exploring the Unknown: Selected Documents in the History of the U.S. Civil Space Program*, vol. 1, *Organizing for Exploration*, ed. John M. Logsdon (Washington, D.C.: NASA History Office, 1995), 379–81.

67. John F. Kennedy to Vice President, April 20, 1961, Presidential Files, JFK Presidential Library, reprinted as document III-6 in Logsdon, *Exploring the Unknown*, 1:424.

68. "Discussion Notes by the Deputy Administrator of the National Aeronautics and Space Administration [Dryden]," April 22, 1961, Vice Presidential File, Space and Space Program, LBJ Presidential Library, reprinted as document 352 in *Foreign Relations of the United States, 1961–63*, vol. 25.

69. Lyndon B. Johnson to John F. Kennedy, "Evaluation of the Space Program," April 28, 1961, NASA Historical Reference Collection, NASA History Office, Washington, D.C., reprinted as Document III-8 in Logsdon, *Exploring the Unknown*, 1:427–28.

70. Newton Minow (head of communications commission) interview, quoted in Robert Dallek, "Johnson, Project Apollo, and the Politics of Space Program Planning," in *Spaceflight and the Myth of Presidential Leadership*, ed. Roger D. Launius and Howard E. McCurdy (Urbana: University of Illinois Press, 1997), 73. On Johnson's interest in international education and satellites when he became president, see Leonard H. Marks to President Lyndon Johnson, memorandum, "Report of White House Working Group on Educational Communications Satellites," August 1966, Cater, Douglass; Working Group on Educational Comm. Satellites (1) folder, box 45, Office Files of White House Aides—S. Douglass Cater, LBJ Presidential Library, Austin, Texas.

71. Overton Brooks to the Vice President, "Recommendations re the National Space Program," May 4, 1961, reprinted in Roger D. Launius, *Apollo: A Retrospective Analysis* (Washington, D.C.: NASA, 1994).

72. John F. Kennedy, "Urgent National Needs," speech to a joint session of Congress, May 25, 1961, NASA Historical Reference Collection, History Office, NASA Headquarters, Washington, D.C., reprinted as document III-12 in Logsdon, *Exploring the Unknown*, 1:454.

73. NASA News Release 61-112, "Statement by James E. Webb, Administrator," May 25, 1961, folder 013923, NASA History Office, Washington, D.C.

74. On NASA keeping track of Hughes's work, see House Subcommittee on Space Sciences of the Committee on Science and Astronautics, *Project Advent– Military Communications Satellite Program: Hearings before the Subcommittee on Space Sciences*, 87th Cong., 2d sess., August 17, 1962, 92. For further details based on interviews with NASA staff, see Delbert D. Smith, *Communication via Satellite: A Vision in Retrospect* (Leyden: Sijthoff, 1976), 84–85.

75. "Inquiry into Problems of Regulating Commercial Space Communication Systems," Public Notice, Report No. 821: Nonbroadcast and General Action, March 30, 1961, FCC, microfilm roll 1, Records of the FCC, JFK Presidential Library, Boston, Mass.

76. House Committee on Science and Astronautics, *Communications Satellites, Part 1*, 51, 73, 75, 135, 142, 145, 310, 408, 413; *Communications Satellites, Part 2*, 87th Cong., 1st sess. (July 14, 17 and August 1, 9–10, 1961), 877, 879. On the military work of the companies, also see David J. Whalen, "Billion Dollar Technology: A Short Historical Overview of the Origins of Communication Satellite Technology, 1945–1965," in *Beyond the Ionosphere: Fifty Years of Satellite Communication*, ed. Andrew J. Butrica (Washington, D.C.: NASA, 1997), 112–22.

77. House Committee on Science and Astronautics, *Communications Satellites, Part 1*, 51, 73, 75, 135, 142, 145, 310, 408, 413; *Communications Satellites, Part 2*, 877, 879.

78. *First Report: In the Matter of an Inquiry into the Administrative and Regulatory Problems Relating to the Authorization of Commercially Operable Space Communications Systems*, May 24, 1961, docket 14024, docket files, Records of the FCC, RG 173, NARA, 2–4.

79. "Inquiry into Problems of Regulating Commercial Space Systems."

80. *First Report*, 1.

81. Testimony of the Department of Justice before the FCC, May 5, 1961, reproduced in House Committee on Interstate and Foreign Commerce, *Communications Satellites, Part 1*, 561–62. On views of the FCC, see *First Report*, 4–5. On the schedule of the Ad Hoc Carrier Committee, see *Supplemental Notice of Inquiry: In the Matter of an Inquiry into the Administrative and Regulatory Problems Relating to the Authorization of Commercially Operable Space Communications Systems*, July 21, 1961, 4, microfilm roll 1, Records of the FCC, JFK Presidential Library, Boston, Mass.

82. On Craven's important role with both the FCC and the TCC, see Robert G. Nunn Jr. to Edward Welsh, April 11, 1961, Nat. Aeronautics and Space Administration folder, box 22, NASC, Gen. Corresp., 1961–69, Records of Temporary Committees, Commission, and Boards, RG 220, NARA. For "Space Commissioner" reference, see interview with Robert E. Lee by Nina Gilden Seavey, July 9, 1985, 7, Comsat History Project, Series III: Comsat History Files, Comsat Corporation Collection, Milton S. Eisenhower Library Special Collections, Johns Hopkins University, Baltimore, Md.

83. *Report of Telecommunications Coordinating Committee Study Group on the Formulation of National Satellite Communication Policy*, May 9, 1961, 3, 5, microfilm roll 1, Records of the FCC, JFK Library, Boston, Mass.

84. *Report of Telecommunications Coordinating Committee Study Group*, 3, 5.

85. Memorandum, "Communications Satellites," E. C. Welsh to David E. Bell, Bureau of the Budget folder, box 14, NASC, Gen. Corresp., 1961–69, Records of Temporary Committees, Commission, and Boards, RG 220, NARA.

86. Memorandum, "U.S. Policy on Communications Satellites," Russell W. Hale to Vice President, June 6, 1961, Nat. Aeronautics and Space Administration folder, box 22, NASC, Gen. Corresp., 1961–69, Records of Temporary Committees, Commission, and Boards, RG 220, NARA. Also see Welsh interview, 4–5.

87. "Chronology of Significant Events: National Aeronautics and Space Council," 9, NASC—History Project folder, box 24, NASC, Gen. Corresp., 1961–69, Records of Temporary Committees, Commission, and Boards, RG 220, NARA.

88. Johnson's instructions were recorded in memorandum for record, June 6, 1961, Communications Satellite Corporation folder, box 15, NASC, Gen. Corresp., 1961–69, Records of Temporary Committees, Commission, and Boards, RG 220, NARA. For a chronology of events leading to the July policy statement on satellite communications, see "Chronology of Significant Events." On the NASC decision leading to the July policy statement, see NASC "Staff Document" on Communication Satellites, July 6, 1961, Science: Space and Aeronautics—Space Council, Communications Satellites folder 1 of 2, box 117, Vice President, 1961–63, Lyndon B. Johnson Library, Austin, Texas.

89. Edward Welsh to Vice President Johnson, June 30, 1961, Communications Satellite Corporation folder, box 15, NASC, Gen. Corresp., 1961–69, Records of Temporary Committees, Commission, and Boards, RG 220, NARA (quotations from attached page "Summary and Highlights: Letter from Harold S. Geneen," June 30, 1961).

90. Robert G. Nunn Jr. quoted in memorandum, "Highlights of Meeting of June 27, 1961, Concerning Communication Satellites," 7, Richard Hirsch to Edward Welsh, June 28, 1961, Staff

Meetings and Drafts, June 5–July 13, 1961, folder, box 14, NASC, Gen. Corresp., 1961–69, Records of Temporary Committees, Commission, and Boards, RG 220, NARA.

91. For "high-level" quotation, see comments of Wreatham Gathright, State Department representative, in memorandum, "Highlights of Meeting of June 9, 1961, Concerning Communications Satellites," Richard Hirsch to Russell Hale, June 9, 1961, Staff Meetings and Drafts, June 5–July 13, 1961, folder, box 14, NASC, Gen. Corresp., 1961–69, Records of Temporary Committees, Commission, and Boards, RG 220, NARA.

92. Memorandum, "Highlights of Meeting of June 12, 1961, Concerning Communications Satellites," Richard Hirsch to Edward C. Welsh, June 13, 1961, Staff Meetings and Drafts, June 5–July 13, 1961, folder, box 14, NASC, Gen. Corresp., 1961–69, Records of Temporary Committees, Commission, and Boards, RG 220, NARA.

93. For Nunn comments, see memorandum, "Highlights of Meeting of June 9, 1961, Concerning Communications Satellites."

94. For statement by Ralph Clark, Defense Department representative, see memorandum, "Highlights of Meeting of June 9, 1961, Concerning Communications Satellites."

95. Seamans quoted in House Subcommittee on Space Sciences, *Project Advent*, 105.

96. House Committee on Interstate and Foreign Commerce, *Communications Satellites, Part 2* (March 21, 1962), 620.

97. Rubel quoted in House Subcommittee on Space Sciences, *Project Advent*, 102.

98. House Subcommittee on Space Sciences, *Project Advent*, 92.

99. *Report of Committee on Department of Defense Communications Satellite Program*, August 1962, 6, Defense—Comsat Corp. Negotiations folder, box 8, Directors Comsat Records 1962–66, Records of the Office of Emergency Preparedness, RG 396, NARA.

100. Smith, *Communications via Satellite*, 85.

101. "Draft from Pentagon Internal Paper," December 1962, Communications Satellite Program (DOD) folder, box 7, Directors Comsat Records 1962–66, Records of the Office of Emergency Preparedness, RG 396, NARA.

102. House Subcommittee on Military Operations of the Committee on Government Operations, *Government Use of Satellite Communications: Hearing before the Subcommittee on Military Operations*, 89th Cong., 2d sess. (September 12, 1966), 526.

103. Senate Communications Subcommittee of the Committee on Commerce, *Space Communications and Allocation of Radio Spectrum: Hearings on S.J. Res. 32*, 87th Cong., 1st sess. (August 24, 1961), 184.

104. House Committee on Interstate and Foreign Commerce, *Communications Satellites, Part 2*, 618–19.

105. For statement by Wreatham Gathright, State Department representative, see memorandum, "Highlights of Meeting of June 9, 1961, Concerning Communications Satellites."

106. "Highlights of Meeting of June 12, 1961, Concerning Communications Satellites."

107. Memorandum, Edward Welsh to Dutton, August 14, 1961, OS2 Aeronautical and Space Vehicle, 1961, folder, box 653, White House Central Subject File, JFK Presidential Library, Boston, Mass.

108. John F. Kennedy to Lyndon Johnson, June 15, 1961, Staff Meetings and Drafts, June 5–July 13, 1961, folder, box 14, NASC, Gen. Corresp., 1961–69, Records of Temporary Committees, Commission, and Boards, RG 220, NARA.

109. "Highlights of Meeting of June 27, 1961, Concerning Communications Satellites."

110. Kennedy to Johnson, June 15, 1961.

111. "Highlights of Meeting of June 12, 1961, Concerning Communications Satellites."

112. Edward C. Welsh to Sterling B. Brinkley, July 6, 1961, OS2 Aeronautical and Space Vehicle, 1961, folder, box 653, White House Central Subject File, JFK Presidential Library, Boston, Mass.

113. Welsh to Brinkley, July 6, 1961.

114. Edward C. Welsh to Overton Brooks, July 18, 1961, Staff Meetings and Drafts, June 5–July 13, 1961, folder, box 14, NASC, Gen. Corresp., 1961–69, Records of Temporary Committees, Commission, and Boards, RG 220, NARA.

115. Memorandum, "Communication Satellite Policy Recommendations," Edward Welsh to President Kennedy, July 15, 1961, Staff Meetings and Drafts, June 5–July 13, 1961, folder, box 14, NASC, Gen. Corresp., 1961–69, Records of Temporary Committees, Commission, and Boards, RG 220, NARA.

116. Memorandum, "Public Version of Space Policy," Philip J. Farley to Edward Welsh, n.d., Busby, Horace, folder, box 14, NASC, Gen. Corresp., 1961–69, Records of Temporary Committees, Commission, and Boards, RG 220, NARA.

117. Memorandum, "Suggestions Concerning Portions of Draft Communications Satellite Policy Paper Which We Do Not Favor Receiving Public Issuance," Henry Geller et al. to Edward Welsh, July 7, 1961, Staff Meetings and Drafts, June 5–July 13, 1961, folder, box 14, NASC, Gen. Corresp., 1961–69, Records of Temporary Committees, Commission, and Boards, RG 220, NARA.

118. Memorandum, Frederick G. Dutton to President John F. Kennedy, July 18, 1961, OS2 Aeronautical and Space Vehicle, 1961, folder, box 653, White House Central Subject File, JFK Presidential Library, Boston, Mass.

119. "Statement of the Department of Justice (before the Federal Communications Commission)," May 5, 1961, reproduced in House Committee on Science and Astronautics, *Communications Satellites, Part 1* (July 13, 1961), 561–62.

120. Memorandum, Dutton to President Kennedy, July 18, 1961.

121. Edward C. Welsh to Vice President Johnson, July 28, 1961, NASC Activities and Accomplishments folder, box 24, NASC, Gen. Corresp., 1961–69, Records of Temporary Committees, Commission, and Boards, RG 220, NARA.

122. Office of the White House Press Secretary, "Statement of the President on Communication Satellite Policy," press release, July 24, 1961, Staff Meetings and Drafts, June 5–July 13, 1961, folder, box 14, NASC, Gen. Corresp., 1961–69, Records of Temporary Committees, Commission, and Boards, RG 220, NARA.

123. Office of the White House Press Secretary, "Statement of the President on Communication Satellite Policy."

124. Lawrence F. O'Brien to Russell B. Long, August 29, 1961, OS2 Aeronautical and Space Vehicle, 1961, folder, box 653, White House Central Subject File, JFK Presidential Library, Boston, Mass.

125. Letter from congressmen is reproduced in Senate Committee on Aeronautical and Space Sciences, *Communication Satellites*, 160–62. Also see Jonathan F. Galloway, *The Politics and Technology of Satellite Communications* (Lexington, Mass.: Lexington Books, 1972), 27, 33–34, 52.

126. *Supplemental Notice of Inquiry: In the Matter of An Inquiry into the Administrative and Regulatory Problems Relating to the Authorization of Commercially Operable Space Communications Systems*, docket No. 14024, July 25, 1961, microfilm roll 1, Records of the FCC, JFK Presidential Library, Boston, Mass.

127. Letter from congressmen reproduced in Senate Committee on Aeronautical and Space Sciences, *Communication Satellites*, 160–62. Also see Galloway, *Politics and Technology of Satellite Communications*, 27, 33–34, 52.

128. Memorandum, Harlan Cleveland to Secretary of State, July 13, 1961, Outer Space—General, 1961, folder, box 94, Subjects, Harlan Cleveland Papers, JFK Presidential Library, Boston, Mass.

129. Cleveland memorandum to Secretary of State, July 13, 1961.

130. Memorandum of conversation, "United Nations Outer Space Committee and Confer-

ence," May 19, 1961, Central Files 1960–63, General Records of the Department of State, RG 59, NARA, reprinted as document 365 in *Foreign Relations of the United States, 1961–1963*, vol. 25.

131. "Ownership: Communication Satellites," September 18, 1961, Communications Satellite Ownership—1961 folder, box 15, NASC, Gen. Corresp., 1961–69, Records of Temporary Committees, Commission, and Boards, RG 220, NARA.

132. "Ownership: Communication Satellites," September 18, 1961.

133. Memorandum, Frederick G. Dutton to Edward Welsh et al., September 22, 1961, Communications Satellite Ownership—1961 folder, box 15, NASC, Gen. Corresp., 1961–69, Records of Temporary Committees, Commission, and Boards, RG 220, NARA.

134. Memorandum, R. W. Hale to Edward Welsh, "Ownership: Communication Satellite," October 4, 1961, Communications Satellite Ownership—1961 folder, box 15, NASC, Gen. Corresp., 1961–69, Records of Temporary Committees, Commission, and Boards, RG 220, NARA.

135. "Synopsis of Agency Letters on Ownership" (no author or date), attached to "Ownership: Communication Satellite," October 4, 1961.

136. Memorandum, R. W. Hale to Edward Welsh, "Ownership: Communication Satellite Staff Choices," October 4, 1961, Communications Satellite Ownership—1961 folder, box 15, NASC, Gen. Corresp., 1961–69, Records of Temporary Committees, Commission, and Boards, RG 220, NARA. Hale quoted here in all cases.

137. Memorandum, Edward C. Welsh to Mr. Dutton, "Consideration of Alternative Proposals for Ownership of Communication Satellite System," October 6, 1961, Communications Satellite Ownership—1961 folder, box 15, NASC, Gen. Corresp., 1961–69, Records of Temporary Committees, Commission, and Boards, RG 220, NARA.

138. Welsh memorandum to Dutton, "Consideration of Alternative Proposals for Ownership of Communication Satellite System."

139. Views of State, Justice, and the Space Council reported by Richard Hirsch, memorandum for Dr. Welsh, "Highlights of Meeting on Monday, 6 November 1961, Concerning Communication Satellite Organization," November 7, 1961, Communications Satellite Ownership—1961 folder, box 15, NASC, Gen. Corresp., 1961–69, Records of Temporary Committees, Commission, and Boards, RG 220, NARA.

140. Loevinger quoted in "Lee Loevinger, 91, Kennedy-Era Antitrust Chief," obituary, *New York Times*, May 8, 2004.

141. Interview with Lee Loevinger by Nina Gilden Seavey, September 18, 1985, 39, 45, Comsat History Project, Series III: Comsat History Files, Comsat Corporation Collection, Milton S. Eisenhower Library Special Collections, Johns Hopkins University, Baltimore, Md.

142. "Nicholas Katzenbach, 90, Dies; Policy Maker at '60s Turning Points," obituary, *New York Times*, May 9, 2012.

143. Views of different agencies as related by Richard Hirsch (his words) in his memorandum for Dr. Welsh, "Highlights of Meeting on Monday, 6 November 1961."

144. Memorandum for the record, R. W. Hale, "Conference on Communication Satellites . . . November 3, 1961," n.d., Communications Satellites Ownership—1961 folder, box 15, NASC, Gen. Corresp., 1961–69, Records of Temporary Committees, Commission, and Boards, RG 220, NARA.

145. Memorandum for the president, Frederick G. Dutton to President Kennedy, November 13, 1961, OS2 Aeronautical and Space Vehicle, 1961, folder, box 653, White House Central Subject File, JFK Presidential Library, Boston, Mass.

146. John Johnson interview, 10–22, quotations on 10, 15, 16, 22.

147. Lyndon B. Johnson to President Kennedy, December 1, 1961, Science, Space and Aeronautics—Space Council, Communications Satellites folder 1 of 2, box 117, Vice President, 1961–63, LBJ Presidential Library, Austin, Texas.

148. Welsh interview, 46–47.

149. Welsh interview, 16.

150. Welsh interview, 17–18. Quotation from interview with Nicholas Katzenbach by Nina Gilden Seavey, September 13, 1985, 4, Comsat History Project, Series III: Comsat History Files, Comsat Corporation Collection, Milton S. Eisenhower Library Special Collections, Johns Hopkins University, Baltimore, Md.

151. Welsh interview, 15.

152. Katzenbach interview, 4–5, quotation on 4.

153. Galloway, *Politics and Technology of Satellite Communications*, 53.

154. Katzenbach interview, 6–8, quotation on 8.

155. Loevinger interview, 6–8, 18, quotation on 18.

156. Loevinger interview, 6–8, 18.

157. Galloway, *Politics and Technology of Satellite Communications*, 50.

158. Interview with Bernard Fensterwald by Nina Gilden Seavey, July 30, 1985, 2, Comsat History Project, Series III: Comsat History Files, Comsat Corporation Collection, Milton S. Eisenhower Library Special Collections, Johns Hopkins University, Baltimore, Md. On Blair and Kefauver, also see Richard E. McFadyen, "Estes Kefauver and the Tradition of Southern Progressivism," *Tennessee Historical Quarterly* 37 (Winter 1978): 430–43.

159. Daniel Scroop, "A Faded Passion? Estes Kefauver and the Senate Subcommittee on Antitrust and Monopoly," *Business and Economic History On-Line* 5 (2007): 1–17, on 10–11.

160. Scroop, "Faded Passion?," 10.

161. Interview of William Gilbert Carter by Nina Gilden Seavey, July 15, 1985, 5–8, Comsat History Project, Series III: Comsat History Files, Comsat Corporation Collection, Milton S. Eisenhower Library Special Collections, Johns Hopkins University, Baltimore, Md.

162. Galloway, *Politics and Technology of Satellite Communications*, 53.

163. Richard G. Hewlett and Jack M. Holl, *Atoms for Peace and War, 1953–1961: Eisenhower and the Atomic Energy Commission* (Berkeley: University of California Press, 1989), 113–43. Also see Asher Ende interview, 5.

164. Galloway, *Politics and Technology of Satellite Communications*, 64–65.

165. Galloway, *Politics and Technology of Satellite Communications*, 65.

166. Galloway, *Politics and Technology of Satellite Communications*, 64–68.

167. Fensterwald interview, 9, 17.

168. Fensterwald interview, 9.

169. Fensterwald interview, 15–16.

170. Galloway, *Politics and Technology of Satellite Communications*, 68–69.

171. Galloway, *Politics and Technology of Satellite Communications*, 69.

172. Fensterwald interview, 8.

173. Fensterwald interview, 9–10.

174. Fensterwald interview, 10–11.

175. William Gilbert Carter interview, 5–9.

176. Peter Temin, *The Fall of the Bell System: A Study in Prices and Politics* (New York: Cambridge University Press, 1987), 28–56.

177. Testimony of Representative Emanuel Celler in Senate Subcommittee on Antitrust and Monopoly of the Committee on the Judiciary, *Antitrust Problems of the Space Satellite Communications System: Hearings on S. R. 258*, pt. 1, 87th Cong, 2d sess. (April 4, 1962), 131.

178. Testimony of Bernard Strassburg (assistant chief, FCC Common Carrier Bureau) in Senate Subcommittee on Monopoly of the Select Committee on Small Business, *Space Satellite Communications: Hearings before the Subcommittee on Monopoly*, 87th Cong., 1st sess. (August 11, 1961), 479–83.

179. Testimony of Lee Loevinger in Senate Subcommittee on Monopoly, *Space Satellite Communications* (August 2, 1961), 42, 52–53.

180. E. L. Hageman (president of the Commercial Telegraphers' Union) to John F. Kennedy, June 11, 1962, Comsat Corp. (Correspondence) folder, box 7, Directors Comsat Records 1962–66, Records of the Office of Emergency Preparedness, RG 396, NARA.

181. Senate Subcommittee on Antitrust and Monopoly, *Antitrust Problems of the Space Satellite Communications System*, pt. 1, 168.

182. Testimony of Senator Wayne Morse in Senate Committee on Foreign Relations, *Communications Satellite Act of 1962: Hearings on H.R. 11040*, 87th Cong., 2d sess. (August 6, 1962), 229.

183. Testimony of Senator Albert Gore in Senate Committee on Foreign Relations, *Communications Satellite Act of 1962* (August 7, 1962), 342.

184. Edward R. Murrow testimony in Senate Committee on Foreign Relations, *Communications Satellite Act of 1962* (August 6, 1962), 128.

185. Testimony of Senator Wayne Morse, in Senate Committee on Foreign Relations, *Communications Satellite Act of 1962* (August 3, 1962), 85.

186. Joseph L. Rauh testimony, quoted in Senate Committee on Foreign Relations, *Communications Satellite Act of 1962*, 229.

187. Testimony of Senator Russell Long in Senate Subcommittee on Monopoly, *Space Satellite Communications*, 401.

188. Adler testimony in Senate Subcommittee on Applications and Tracking and Data Acquisition, *Commercial Communications Satellites: Hearings on S. R. 258*, pt. 1, 87th Cong., 2d sess. (1962), 7, 25. On Senate staff members echoing Hughes's arguments, see testimony of Bernard Fensterwald (staff director), April 12, 1962, in Senate Subcommittee on Antitrust and Monopoly, *Antitrust Problems of the Space Satellite Communications System*, pt. 2, 437.

Chapter 3 · *Global Satellite Communications and the 1963 International Telecommunication Union Space Radio Conference*

1. Memorandum for the president, "Appointment of a Special Assistant to the President for Telecommunications," March 11, 1964, ND 3 Communications—Electronics folder, box 4, National Security—Defense ND 3, Lyndon B. Johnson (LBJ) Presidential Library, Austin, Texas.

2. "Department of Defense/Communications Satellite Corporation Agreement," June 29, 1964, Defense—COMSAT Corp. Negotiations folder, box 8, Directors COMSAT Records 1962–66, Records of the Office of Emergency Planning, Record Group (RG) 396, National Archives and Records Administration (NARA), College Park, Md.

3. James Schwoch, *Global TV: New Media and the Cold War, 1946–69* (Urbana: University of Illinois Press, 2009), 124.

4. House Subcommittee of the Committee on Government Operations, *Satellite Communications—1964 (Part I): Hearings*, 88th Congress, 2d sess. (March 19 and April 15, 1964), 63, 353. The Orrick Committee was officially the Subcommittee on Communications of the Executive Committee of the National Security Council. On the establishment of the Orrick Committee, see McGeorge Bundy to Vice President et al., "Establishment of Subcommittee on Communications," October 26, 1962, S/S-NSC Files, Lot 72 D 316, NSAM 201, General Records of the Department of State, RG 59, NARA, reprinted as document 438 in U.S. State Department, *Foreign Relations of the United States, 1961–63*, vol. 25, *Organization of Foreign Policy, Information Policy, United Nations, Scientific Matters, Communication Satellites*.

5. Memorandum from the chairman of the Subcommittee on Communications of the National Security Council (Orrick) to the Executive Committee of the National Security Council, May 21, 1963, NSAM No. 201, box 339, Meetings and Memoranda Series, National Security Files, John F. Kennedy Library, reprinted as document 442 in U.S. State Department, *Foreign Relations of the United States, 1961–63*, vol. 25.

6. A special subcommittee of the Orrick Committee was established on January 28, 1963, to

evaluate the role of satellite communications in the proposed National Communications System. W. Michael Blumenthal was appointed chairman. See memorandum from Orrick to Department of State Principals, January 28, 1963, IO Files, Lot 67 D 378, ITU, General Records of the Department of State, RG 59, NARA, reprinted as document 433 in U.S. State Department, *Foreign Relations of the United States, 1961–63*, vol. 25. Other important material on the Orrick Committee is located in box 441, National Security Files, John F. Kennedy (JFK) Presidential Library, Boston, Mass. On the advantages of satellite communications in the event of a nuclear attack, see testimony of Lt. Gen. Alfred D. Starbird in House Subcommittee of the Committee on Government Operations, *Satellite Communications—1964 (Part I)* (March 19, 1964), 77. This was the only time that the National Security Council dealt with satellite communications. During this period, the National Security Council did not deal with outer space policy. President Kennedy delegated space policy to the National Aeronautics and Space Council. See memorandum from the counselor of the Department of State and chairman of the Policy Planning Council (McGhee) to the president's special assistant for national security affairs (Bundy), March 28, 1961, Policy Planning, 2/11/61–5/61, Subject Series, National Security Files, JFK Presidential Library, reprinted as document 10 in US State Department, *Foreign Relations of the United States, 1961–63*, vol. 25.

7. House Subcommittee of the Committee on Government Operations, *Satellite Communications—1964 (Part I)* (April 15, 1964), 354.

8. For "90 payloads" quotation, see E. C. Welsh, memorandum to the vice president, "Military vs Non-Military Space Activities," January 19, 1963, DEFENSE-1963 folder, box 17, NASC, Gen. Corresp., 1961–69, Records of Temporary Committees, Commission, and Boards, RG 220, NARA. For "two-thirds" quotation, see "National Aeronautics and Space Council Meeting— September 28, 1962," box 3, NASC, Gen. Corresp., 1961–69, Records of Temporary Committees, Commission, and Boards, RG 220, NARA.

9. Testimony of Joseph V. Charyk in House Subcommittee of the Committee on Government Operations, *Satellite Communications—1964 (Part I)* (March 24, 1964), 109.

10. Senate Communications Subcommittee of the Committee on Commerce, *Space Communications and Allocation of Radio Spectrum: Hearings on Space Communications and S.J. Res. 32* (August 24, 1961), 183.

11. "Industry-Department of Defense Cooperation in Satellite-Based Telecommunications," September 7, 1961, unmarked first folder, box 7, Directors COMSAT Records 1962–66, Records of the Office of Emergency Planning, RG 396, NARA.

12. Brown testimony in Senate Communications Subcommittee, *Space Communications and Allocation of Radio Spectrum* (August 24, 1961), 184.

13. John H. Rubel (assistant secretary of defense deputy director, Defense Research and Engineering), "Gray Paper on the Subject of Advent Communication Satellite System," April 3, 1962, Communications Satellites folder, box 7, Directors COMSAT Records 1962–66, Records of the Office of Emergency Planning, RG 396, NARA.

14. Testimony of Eugene G. Fubini in House Subcommittee of the Committee on Government Operations, *Satellite Communications—1964 (Part I)* (March 18, 1964), 42. For "overruns" quotation, see "Draft from Pentagon Internal Paper," white paper on the DOD Communications Satellite Program, January 1963, Communications Satellite Program (DOD) folder, box 7, Directors COMSAT Records 1962–66, Records of the Office of Emergency Planning, RG 396, NARA.

15. "Draft from Pentagon Internal Paper." Also see editorial by Victor de Biasi, "Another Go-Around," *Space/Astronautics*, October 1964, copy in COMSAT—Miscellaneous folder, box 7, Directors COMSAT Records 1962–66, Records of the Office of Emergency Planning, RG 396, NARA; and Irvin Stewart to Oren Harris, March 15, 1963, unmarked first folder, box 7, Directors COMSAT Records 1962–66, Records of the Office of Emergency Planning, RG 396, NARA.

16. Nicholas deB. Katzenbach, memorandum for the president, May 6, 1963, UT1: Communications—Telecommunications folder, box 96, White House Central Files Confidential File, LBJ Presidential Library, Austin, Texas. This was part of an effort by McNamara during this period to eliminate "duplicative and wasteful programs." See Dwayne A. Day, "Invitation to Struggle: The History of Civilian-Military Relations in Space," in *Exploring the Unknown: Selected Documents in the History of the U.S. Civilian Space Program*, vol. 2, *External Relationships*, ed. John M. Logsdon (Washington, D.C.: NASA History Office, 1996), 260.

17. Robert Kennedy (attorney general) to President Kennedy, October 25, 1962, and minutes of first meeting of COMSAT Board of Directors (incorporators), February 4, 1963, both in bound book of documents, *Communications Satellite Corporation—Establishment of the Corporation*, box 4, Beardsley Graham Paper, Bancroft Library, University of California, Berkeley. President Kennedy appointed the incorporators on "recess basis" because the "Senate in its closing days was unable to act on nominations." See President Kennedy to "Dear Mr.," October 15, 1962, in *Communications Satellite Corporation—Establishment of the Corporation*.

18. President Kennedy to "Dear Mr.," October 15, 1962.

19. Interview with Joseph Charyk by Nina Gilden Seavey, April 1, 1986, 6, Comsat History Project, Series III: Comsat History Files, Comsat Corporation Collection, Milton S. Eisenhower Library Special Collections, Johns Hopkins University, Baltimore, Md. Also see similar comments by another early employee, Leonard Marks: interview with Leonard Marks by Nina Gilden Seavey, August 9, 1985, 4, Comsat History Project, Series III: Comsat History Files, Comsat Corporation Collection, Milton S. Eisenhower Library Special Collections, Johns Hopkins University, Baltimore, Md.

20. Marks interview, 7.

21. On role of Lloyd Cutler, see interview with William Berman by Thomas Maxwell Safely, December 10, 1984, 13, Comsat History Project, Series III: Comsat History Files, Comsat Corporation Collection, Milton S. Eisenhower Library Special Collections, Johns Hopkins University, Baltimore, Md.

22. Prospectus, "Communications Satellite Corporation," June 1964, 28, folder 4, box 35, Joseph V. Charyk Papers, Special Collections Research Center, George Washington University, Washington, D.C. For quotation about Harris, see interview with Allen Throop by Thomas Maxwell Safely, July 18, 1984, 4, Comsat History Project, Series III: Comsat History Files, Comsat Corporation Collection, Milton S. Eisenhower Library Special Collections, Johns Hopkins University, Baltimore, Md. On Marks, see Marks interview, 9.

23. Throop interview, 3.

24. Throop interview, 7.

25. Charyk interview, 84–85. Also see "Minutes of First Meeting of Board of Directors," February 4, 1963, in *Communications Satellite Corporation—Establishment of the Corporation*.

26. Marks interview, 3.

27. Charyk interview, 5.

28. Charyk interview, 13–14.

29. Marks interview, 8.

30. Throop interview, 1–2; Charyk interview, 44; interview with David Melamed by Nina Gilden Seavey, August 20, 1985, 1, Comsat History Project, Series III: Comsat History Files, Comsat Corporation Collection, Milton S. Eisenhower Library Special Collections, Johns Hopkins University, Baltimore, Md.

31. Interview with Donald Greer by Thomas Maxwell Safely, December 10, 1984, 22–23, Comsat History Project, Series III: Comsat History Files, Comsat Corporation Collection, Milton S. Eisenhower Library Special Collections, Johns Hopkins University, Baltimore, Md.

32. Charyk interview, 40–41.

33. Interview with Sidney Metzger by Thomas Maxwell Safely, July 11, 1984, 4–5, Comsat

History Project, Series III: Comsat History Files, Comsat Corporation Collection, Milton S. Eisenhower Library Special Collections, Johns Hopkins University, Baltimore, Md. For a discussion of early Comsat employees, see David J. Whalen, *The Rise and Fall of Comsat: Technology, Business, and Government in Satellite Communications* (Basingstoke: Palgrave Macmillan, 2014), 37–45, 131–34.

34. Interview with Lewis Meyer by Thomas Maxwell Safely, July 24, 1984, 1, Comsat History Project, Series III: Comsat History Files, Comsat Corporation Collection, Milton S. Eisenhower Library Special Collections, Johns Hopkins University, Baltimore, Md.; and Greer interview, 1.

35. Charyk interview, 46; interview with John Johnson by Nina Gilden Seavey, March 13, 1986, 1–2, Comsat History Project, Series III: Comsat History Files, Comsat Corporation Collection, Milton S. Eisenhower Library Special Collections, Johns Hopkins University, Baltimore, Md. For quotations, see Berman interview, 51–52.

36. Greer interview, 10.

37. Interview with James Potts by Thomas Maxwell Safely, July 24, 1984, 3, Comsat History Project, Series III: Comsat History Files, Comsat Corporation Collection, Milton S. Eisenhower Library Special Collections, Johns Hopkins University, Baltimore, Md.

38. Interview with Matthew Gordon by Nina Gilden Seavey, September 11, 1985, 1, Comsat History Project, Series III: Comsat History Files, Comsat Corporation Collection, Milton S. Eisenhower Library Special Collections, Johns Hopkins University, Baltimore, Md.

39. Interview with Bill Callaway, n.d., Comsat History Files, n.p., Series III: Comsat History Files, Comsat Corporation Collection, Milton S. Eisenhower Library Special Collections, Johns Hopkins University, Baltimore, Md.

40. Greer interview, 2.

41. Charyk interview, 35.

42. Charyk interview, 37.

43. Throop interview, 15.

44. Throop interview, 15.

45. Charyk interview, 27.

46. Charyk interview, 28.

47. John Johnson interview, 67.

48. Charyk interview, 33.

49. John Johnson interview, 62.

50. Interview with Asher Ende by Nina Gilden Seavey, August 28, 1985, 30, Comsat History Project, Series III: Comsat History Files, Comsat Corporation Collection, Milton S. Eisenhower Library Special Collections, Johns Hopkins University, Baltimore, Md.

51. Prospectus, "Communications Satellite Corporation."

52. Throop interview, 3.

53. Vartanig G. Vartan, "Public Snaps Up Satellite Corp.'s Shares," *New York Times*, June 3, 1964, file 010718, NASA History Office, Washington, D.C.

54. Katherine Johnsen, "Public Sale Completes Comsat Financing," *Aviation Week and Space Technology*, June 8, 1964, 19, file 010718, NASA History Office, Washington, D.C.

55. Throop quoted in Ende interview, 30.

56. For quotations, see Charyk interview, 31, 33. Also see John Johnson interview, 68.

57. John P. MacKenzie, "Industry Snaps Up Its Half of Comsat Stock," *Washington Post*, May 28, 1964, file 01078, NASA History Office, Washington, D.C.

58. "Report on First Annual Meeting of Shareholders and Recent Developments: Communications Satellite Corporation," file 010718, NASA History Office, Washington, D.C.

59. "CSC Notifies FCC of Intention to Award Global Communications Satellite Design Contracts," Communications Satellite Corporation news release, June 8, 1964, 1, 7, on 7, file 010718, NASA History Office, Washington, D.C.

60. Harold Brown to Nicholas deB. Katzenbach, April 23, 1963, folder 3, box 35, MS2137, Joseph V. Charyk Papers, Special Collections Research Center, George Washington University, Washington, D.C.

61. Memorandum, Fred C. Alexander to all federal agencies, August 1, 1960, reprinted in Senate Communications Subcommittee, *Space Communications and Allocation of Radio Spectrum* (August 23, 1961), 136–37.

62. Letter from Fred Alexander to Paul D. Miles, executive secretary of the IRAC, November 10, 1960, reprinted in Senate Communications Subcommittee, *Space Communications and Allocation of Radio Spectrum* (August 23, 1961), 139.

63. Maurice Edwin Curts testimony in Senate Communications Subcommittee, *Space Communications and Allocation of Radio Spectrum* (August 24, 1961), 185.

64. Fred Alexander testimony in Senate Communications Subcommittee, *Space Communications and Allocation of Radio Spectrum* (August 23, 1961), 153, 155.

65. The president established the position of director of telecommunications management with Executive Order 10995. The director of telecommunications management also served as one of the assistant directors of the Office of Emergency Planning. See National Security Action Memorandum No. 252/1, n.d., S/S-NSC Files, Lot 72 D 316, General Records of the Department of State, RG 59, NARA, reprinted as document 444 in U.S. Department of State, *Foreign Relations of the United States, 1961–63*, vol. 25.

66. House Subcommittee of the Committee on Government Operations, *Satellite Communications—1964 (Part I)* (April 9, 1964), 291–92.

67. Robert G. Nunn Jr. to Edward Welsh, April 11, 1961, box 22, National Aeronautics and Space Administration folder, NASC, Gen. Corresp., 1961–69, Records of Temporary Committees, Commission, and Boards, RG 220, NARA.

68. Senate Committee on Aeronautical and Space Sciences, *Communication Satellites: Technical, Economic, and International Developments*, staff report prepared by Donald R. MacQuivey, 87th Cong., 2d sess. (1962), Committee Print (CIS-NO: S0525), 84; "Preliminary Views of the United States of America: Frequency Allocations for Space Radio Communications," May 17, 1961, box 821, docket 13522, docket files, Records of the Federal Communications Commission (FCC), RG 173, NARA.

69. "Draft Proposals of the United States of America for the Extraordinary Administrative Radio Conference for Space Radio Communication (Geneva, 1963)," October 5, 1962, 9, folder 399.40-GE/8-962, box 842, Central Decimal File 1960–63, General Records of the Department of State, RG 59, NARA.

70. G. Griffith Johnson to Edward A. Bolster, January 4, 1963, folder 399.40-GE/11-2062, box 842, Central Decimal File 1960–63, General Records of the Department of State, RG 59, NARA.

71. "Summary Outline of US Frequency Allocation Proposals and Limitations for Space Users," October 1, 1962, COMSATS—Miscellaneous folder, box 7, Directors COMSAT Records 1962–66, Records of the Office of Emergency Planning, RG 396, NARA.

72. Johnson to Bolster, January 4, 1963. Frequencies for space technology included in early US proposals included the following: 136–174 MHz and 406–470 MHz for Earth-to-space satellite command purposes; 1435–1660 MHz for telemetry and command purposes; 1660–1700 MHz for meteorological satellites; 1700–2300 MHz for deep space research and command; 3.7–4.2 GHz and 5.925–8.4 for communication satellites; 8.4–8.5 GHz for space research; 9.8–10.0 GHz for radiolocation; 15.15–15.25 GHz and 31.5–31.8 GHz for space research; and 33.4–36.0 GHz for radiolocation. See "Preliminary Views of the United States of America."

73. Memorandum, Charles E. Bohlen to Mr. Hare, November 22, 1960, folder 399.40/10-360, box 841, Central Decimal File 1960–1963, General Records of the Department of State, RG 59, NARA.

74. Report by John P. Hagen (chairman of US delegation), meeting of CCIR Study Group IV—Space Systems, May 23, 1962, folder 399.40/5–362, box 841, Central Decimal File 1960–63, General Records of the Department of State, RG 59, NARA.

75. Draft of letter by Edwin M. Martin to industry, September 15, 1962, folder 399.20-ITU/9–161, box 838, Central Decimal File 1960–63, General Records of the Department of State, RG 59, NARA.

76. Summary record, "Washington Arrangements Committee for CCIR Study Groups IV and VIII," September 19, 1961, folder 399.20-ITU/9–161, box 838, Central Decimal File 1960–63, General Records of the Department of State, RG 59, NARA.

77. Report by John Hagen, meeting of CCIR Study Group IV, May 23, 1962.

78. Report by John Hagen, meeting of CCIR Study Group IV, May 23, 1962.

79. Senate Committee on Aeronautical and Space Sciences, *Communication Satellites*, 126.

80. Senate Committee on Aeronautical and Space Sciences, *Communication Satellites*, 150.

81. Francis Colt de Wolf to Fred C. Alexander, January 19, 1961, folder 399.40/8–361, box 841, Central Decimal File 1960–63, General Records of the Department of State, RG 59, NARA.

82. Paul D. Miles to Francis Colt de Wolf, February 13, 1961, folder 399.40/8–361, box 841, Central Decimal File 1960–63, General Records of the Department of State, RG 59, NARA.

83. Instruction to embassies, Department of State, April 17, 1962, folder 399.40/3–262, box 841, Central Decimal File 1960–63, General Records of the Department of State, RG 59, NARA.

84. Memorandum, chairman of TCC Ad Hoc Working Group to principal and alternate members, May 1, 1961, microfilm roll 1, Records of the FCC, JFK Presidential Library, Boston, Mass.

85. Walt Whitman Rostow, *The Stages of Economic Growth: A Non-Communist Manifesto* (Cambridge: Cambridge University Press, 1960); David Halberstam, *The Best and the Brightest* (New York: Ballantine, 1992), 123; Michael E. Latham, *Modernization as Ideology: American Social Science and "Nation Building" in the Kennedy Era* (Chicago: University of Chicago Press, 2000).

86. Department of State, Leopoldville (Congo), to Washington, D.C., August 23, 1960, folder 399.20-ITU/9–160, box 817, Central Decimal File 1960–63, General Records of the Department of State, RG 59, NARA. Also see Christiane Berth, "ITU, The Development Debate, and Technical Cooperation in the Global South, 1950–1992," in *History of the International Telecommunication Union*, ed. Gabriele Balbi and Andreas Fickers (Berlin: De Gruyter Oldenbourg, 2020), 71–76. Soviet involvement in UN assistance programs did increase during the 1950s, but even in the early 1960s, it was still limited. See Robert G. Weston, "The United Nations in the World Outlook of the Soviet Union and the United States," in *Soviet and American Policies in the United Nations: A Twenty-Five Year Perspective*, ed. Alvin Z. Rubinstein and George Ginsburgs (New York: New York University Press, 1971), 15.

87. C. W. Loeber to William G. Carter, August 15, 1962, folder 399.20-ITU/7–162, box 839, Central Decimal File 1960–63, General Records of the Department of State, RG 59, NARA.

88. John Hagen to Irvin Stewart, July 18, 1962, folder 399.20-ITU/7–162, box 839, Central Decimal File 1960–63, General Records of the Department of State, RG 59, NARA.

89. Irvin Stewart to G. Griffith Johnson, August 13, 1962, folder 399.20-ITU/7–162, box 839, Central Decimal File 1960–63, General Records of the Department of State, RG 59, NARA.

90. For quotations and discussion of Stewart's recommendations, see Loeber to Carter, August 15, 1962. Also see Stewart to Johnson, August 13, 1962.

91. G. Griffith Johnson to Irvin Stewart, August 24, 1962, folder 399.20-ITU/7–162, box 839, Central Decimal File 1960–63, General Records of the Department of State, RG 59, NARA.

92. Johnson to Stewart, August 24, 1962.

93. Department of State to Bogota embassy, April 26, 1963, TEL 10, Telegraph, 2/1/63, folder, box 3656, Central Foreign Policy File 1963, General Records of the Department of State, RG 59,

NARA; Fred C. Alexander to G. Griffith Johnson, August 2, 1963, TEL 8, Radio, 2/1/63, folder, box 3655, Central Foreign Policy File 1963, General Records of the Department of State, RG 59, NARA.

94. C. R. Kirkevold to Fred C. Alexander, February 28, 1963, folder 399.20-ITU/5–162, box 839, Central Decimal File, 1960–63, General Records of the Department of State, RG 59, NARA.

95. Department of State to European embassies, January 15, 1963, folder 399.20-ITU/11–162, box 838, Central Decimal File 1960–63, General Records of the Department of State, RG 59, NARA.

96. For State Department instructions to embassies, see "Participation in the ITU Space Radio Communication Conference," September 3, 1963, TEL-6-1, Space Communication Frequencies, 2/1/63, ITU, folder, box 3655, Central Foreign Policy File 1963, General Records of the Department of State, RG 59, NARA; and "Composition of United States Delegation to the Space Radio Conference," September 9, 1963, TEL-6-1, Space Communication Frequencies, 2/1/63, ITU, folder, box 3655, Central Foreign Policy File 1963, General Records of the Department of State, RG 59, NARA. For examples of embassies informing the State Department about the responses of foreign governments to their inquiries, see incoming telegrams in boxes 3654 and 3655, Central Foreign Policy File 1963, General Records of the Department of State, RG 59, NARA.

97. William H. Watkins to Dean Rusk, August 9, 1962, folder 399.40-GE/8–962, box 839, Central Decimal File 1960–63, General Records of the Department of State, RG 59, NARA.

98. William G. Carter to Max Isenbergh, January 8, 1963, folder 399.40-GE/11–2062, box 842, Central Decimal File 1960–63, General Records of the Department of State, RG 59, NARA.

99. Kirkevold to Alexander, February 28, 1963. On the seven-country conference, see memorandum, "Proposed United States Delegations to the . . . Meeting of Experts . . . ," February 26, 1963, TEL 6-1, Space Communication Frequencies, 2/1/63, Frequencies folder, box 3654, Central Foreign Policy File, 1963, General Records of the Department of State, RG 59, NARA.

100. "Proposals for the United States of America for the Extraordinary Administrative Radio Conference for Space Radiocommunication and Radio Astronomy," folder 399.20-ITU/5–162, box 839, Central Decimal File 1960–63, General Records of the Department of State, RG 59, NARA.

101. Johnson to Bolster, January 4, 1963. For frequency comparison, see Department of State to Austrian embassy, March 18, 1963, TEL 8, Radio, 7/1/63, folder, box 3655, Central Foreign Policy File 1963, General Records of the Department of State, RG 59, NARA.

102. Department of State to Austrian embassy, March 18, 1963. Relevant details involving the Soviet space program remain classified in closed archives. See Asif A. Siddiqi, *Sputnik and the Soviet Space Challenge* (Gainesville: University Press of Florida, 2000), 517.

103. Telegram, Austrian embassy to the Department of State, August 16, 1968, SP 6 UN, Central Files, 1967–69, General Records of the Department of State, RG 59, NARA, reprinted as document 104 in U.S. State Department, *Foreign Relations of the United States, 1964–1968*, vol. 34, *Energy, Diplomacy, and Global Issues, Communication Satellites*.

104. Burton I. Edelson (NASC staff), memorandum for the record, "Soviet Communications Satellite Effort," November 19, 1962, COMSAT International Agreement folder, box 8, Directors COMSAT Records 1962–66, Records of the Office of Emergency Planning, RG 396, NARA (this box can be found at the following stack location: 650/86/08/4).

105. Untitled document, n.d., folder 399.20-ITU/5–162, box 839, Central Decimal File 1960–63, General Records of the Department of State, RG 59, NARA.

106. Richard N. Gardner to assistant secretary of the Department of State, September 19, 1963, Telecommunications: TEL 3, Organizations and Conferences, 9/1/63, ITU, folder box 3652, Central Foreign Policy File 1963, General Records of the Department of State, RG 59, NARA. On boundary work, see especially Hugh R. Slotten, *Radio and Television Regulation: Broadcast Technology in the United States, 1920–1960* (Baltimore, Md.: Johns Hopkins University, 2000).

107. Lyle Moore to Pierre Salinger, January 16, 1962, folder 399.40-GE/8–962, box 842, Central Decimal File 1960–63, General Records of the Department of State, RG 59, NARA.

108. Richard S. Wheeler to Thomas N. Gautier, April 3, 1961, folder 399.20-ITU/4–161, box 838, Central Decimal File 1960–63, General Records of the Department of State, RG 59, NARA.

109. Bogota embassy to Department of State, June 10, 1963, Telecommunications: TEL 3, Organizations and Conferences, 4/1/63, ITU, folder, box 3653, Central Foreign Policy File 1963, General Records of the Department of State, RG 59, NARA.

110. Harlan Cleveland to J. Herbert Holloman, May 27, 1963, Telecommunications: TEL 3, Organizations and Conferences, 4/1/63, ITU, folder, box 3653, Central Foreign Policy File 1963, General Records of the Department of State, RG 59, NARA.

111. R. R. Hough to Francis Cunningham, July 3, 1963, TEL 8-1, Radio Frequencies, 2/1/63, ITU, folder, box 3655, Central Foreign Policy File 1963, General Records of the Department of State, RG 59, NARA.

112. Francis O. Wilcox to Gerald C. Gross, January 6, 1960, folder 399.20-ITU/1–160, box 837, Central Decimal File 1960–63, General Records of the Department of State, RG 59, NARA.

113. Roger W. Tubby to Dean Rusk, July 22, 1963, Telecommunications: TEL 3, Organizations and Conferences, 4/1/63, ITU, folder, box 3652, Central Foreign Policy File 1963, General Records of the Department of State, RG 59, NARA.

114. "Classified Report of the US Representative on the Administrative Council of the ITU," 18th Sess., Geneva, March 23–April 26, 1963, 2, Telecommunications: TEL 3, Organizations and Conferences, 4/1/63, ITU, folder, box 3653, Central Foreign Policy File 1963, General Records of the Department of State, RG 59, NARA.

115. Department of State to foreign embassies, May 24, 1963, TEL 8, Radio, 7/1/63, folder, box 3655, Central Foreign Policy File 1963, General Records of the Department of State, RG 59, NARA.

116. Cleveland to Holloman, May 27, 1963.

117. Francis Colt de Wolf to F. R. Cappel, September 6, 1960, folder 399.40-GE/3–260, box 842, Central Decimal File 1960–63, General Records of the Department of State, RG 59, NARA.

118. Frank B. Ellis to Chester Bowles, June 5, 1961, folder 399.40-GE/3–260, box 842, Central Decimal File 1960–63, General Records of the Department of State, RG 59, NARA.

119. De Wolf to Alexander, January 19, 1961.

120. Irvin Stewart to John A. Morrison, March 6, 1963, TEL 6-1, Space Communication Frequencies, 2/1/63, Frequencies, folder, box 3655, Central Foreign Policy File, 1963, General Records of the Department of State, RG 59, NARA.

121. Department of State to foreign embassies, March 10, 1963, TEL 6-1, Space Communication Frequencies, 2/1/63, Frequencies, folder, box 3654, Central Foreign Policy File, 1963, General Records of the Department of State, RG 59, NARA.

122. "Report of the United States Delegation to the Tenth Plenary Assembly of the ITU International Radio Consultative Committee," March 15, 1963, 20, 28, 37, TEL 6, 3/1/73 [sic], folder, box 3654, Central Foreign Policy File, 1963, General Records of the Department of State, RG 59, NARA.

123. Joseph H. McConnell to George W. Ball, October 23, 1963, TEL 6-1, Space Communication Frequencies, 10/1/63, ITU, folder, box 3654, Central Foreign Policy File, 1963, General Records of the Department of State, RG 59, NARA.

124. Geneva to Secretary of State, November 6, 1963, TEL 6-1, Space Communication Frequencies, 10/1/63, ITU, folder, box 3654, Central Foreign Policy File, 1963, General Records of the Department of State, RG 59, NARA.

125. "Classified Report of the United States Territories Delegation to the EARC," November 20, 1963, 3, Telecommunications: TEL 8-1, Radio Frequencies, 8/1/63, ITU, folder, box 3655, Central Foreign Policy File 1963, General Records of the Department of State, RG 59, NARA. The

United States Territories delegation was separate from the main US delegation but still under McConnell's authority.

126. W. Dean to Captain Raish, October 25, 1963, TEL 6, 6/1/63, ITU, folder, box 3654, Central Foreign Policy File, 1963, General Records of the Department of State, RG 59, NARA.

127. "Classified Report of the United States Territories Delegation to the EARC," 3.

128. William Taubman, *Khrushchev: The Man and His Era* (New York: Norton, 2003), 583; David Tal, *The American Nuclear Disarmament Dilemma, 1945–1963* (Syracuse, N.Y.: Syracuse University Press, 2008), 229; Vladislav M. Zubok, *A Failed Empire: The Soviet Union in the Cold War from Stalin to Gorbachev* (Chapel Hill: University of North Carolina Press, 2007), 150.

129. Aleksandr Fursenko and Timothy Naftali, *Khrushchev's Cold War: The Inside Story of an American Adversary* (New York: Norton, 2006), 520.

130. Taubman, *Khrushchev*, 585, 607; Fursenko and Naftali, *Khrushchev's Cold War*, 513.

131. Matthew J. Von Bencke, *The Politics of Space: A History of U.S.-Soviet/Russian Competition and Cooperation in Space* (Boulder, Colo.: Westview Press, 1997), 71; Fursenko and Naftali, *Khrushchev's Cold War*, 520–21.

132. Gordon S. Barrass, *The Great Cold War: A Journey through the Hall of Mirrors* (Stanford, Calif.: Stanford University Press, 2009), 145; Von Bencke, *Politics of Space*, 71; Fursenko and Naftali, *Khrushchev's Cold War*, 525; Zubok, *Failed Empire*, 152.

133. Walter LaFeber, *America, Russia, and the Cold War, 1945–2006* (New York: McGraw Hill, 2008), 237; Benjamin P. Greene, *Eisenhower, Science Advice, and the Nuclear Test-Ban Debate, 1945–1963* (Stanford, Calif.: Stanford University Press, 2007), 239.

134. Wilfried Loth, *Overcoming the Cold War: A History of Détente, 1950–1991* (Basingstoke: Palgrave, 2002), 77.

135. Editorial note on memorandum of understanding in U.S. Department of State, *Foreign Relations of the United States, 1961–63*, vol. 25, document 400.

136. On Dryden-Blagonravov talks, see documents I-36 through I-40 in Logsdon, *Exploring the Unknown*, 2:143–64.

137. Taubman, *Khrushchev*, 605.

138. "Qualifications of the Proposed United States Delegation to the Extraordinary Administrative Radio Conference," Telecommunications: TEL 8-1, Radio Frequencies, 8/1/63, ITU, folder, box 3655, Central Foreign Policy File 1963, General Records of the Department of State, RG 59, NARA. On the final membership list of the delegation, see "Composition of United States Delegation to the Space Radio Conference," October 7, 1963, TEL 6, 6/1/63 folder, box 3654, Central Foreign Policy File, 1963, General Records of the Department of State, RG 59, NARA. On involvement of members of Congress, see "Report on Geneva Space Radio Communication Conference and Progress made in Establishing Global Communication Satellite System," *Congressional Record*, 88th Cong., 2d sess., vol. 110 (January 9, 1964), 181.

139. "Opening of the Conference on Space Radiocommunications: 70 Countries Take Part," *Telecommunication Journal* 30, no. 11 (1963): 334–36, quotation on 335, http://search.itu.int /history/HistoryDigitalCollectionDocLibrary/4.89.57.en.104.pdf.

140. Interview with William Gilbert Carter by Nina Gilden Seavey, July 15, 1985, 46, Comsat History Project, Series III: Comsat History Files, Comsat Corporation Collection, Milton S. Eisenhower Library Special Collections, Johns Hopkins University, Baltimore, Md.

141. Alexander to Johnson, August 2, 1963.

142. "Report on Geneva Space Radio Communication Conference," 182.

143. Burton I. Edelson (NASC staff), memorandum, "Soviet Communications Satellite Effort."

144. McConnell to Ball, October 23, 1963.

145. "Report on Geneva Space Radio Communication Conference," 174.

146. From speech in Congress by Congressman Oren Harris in "Report on Geneva Space Radio Communication Conference," 178.

147. During the press conference using the *Syncom II* satellite, a State Department official did indicate that "some" of the frequencies set aside for satellite communications would be used by the military system being developed. See "Report on Geneva Space Radio Communication Conference," 179.

148. Deputy secretary of defense to U. Alexis Johnson, April 16, 1963, DEFENSE—1963 folder, box 17, NASC, Gen. Corresp., 1961–69, Records of Temporary Committees, Commission, and Boards, RG 220, NARA.

149. G. Griffith Johnson to Harold Brown, October 10, 1963, TEL 6-1, Space Communication Frequencies, 10/1/63, ITU, folder, box 3654, Central Foreign Policy File, 1963, General Records of the Department of State, RG 59, NARA.

150. Telegram, Geneva to secretary of state, October 23, 1963, TEL 6-1, Space Communication Frequencies, 10/1/63, ITU, folder, box 3654, Central Foreign Policy File, 1963, General Records of the Department of State, RG 59, NARA.

151. "Report on Geneva Space Radio Communication Conference," 183. For second quotation, see "Additional Position Papers for the United States Delegation to the Extraordinary Administrative Radio Conference for Space Radiocommunication and Radio Astronomy (Geneva, 1963)," August 23, 1963, agenda item 12a, 1, TEL 6-1, Space Communication Frequencies, 2/1/63, ITU, folder, box 3655, Central Foreign Policy File, 1963, General Records of the Department of State, RG 59, NARA.

152. McConnell to Ball, October 23, 1963.

153. "Department of Defense/Communications Satellite Corporation Agreement."

154. Interview with Joseph McConnell by Frederick C. Durant III, July 18, 1985, 4, Comsat History Project, Series III: Comsat History Files, Comsat Corporation Collection, Milton S. Eisenhower Library Special Collections, Johns Hopkins University, Baltimore, Md.

155. Joseph Charyk, daily log, October 24, 1963, folder 9, box 25, Joseph V. Charyk Papers, Special Collections Research Center, George Washington University, Washington, D.C.

156. Charyk interview.

157. McConnell interview, 4.

158. McConnell interview, 3.

159. McConnell interview, 6.

160. Carter interview.

161. Senate Communications Subcommittee, *Space Communications and Allocation of Radio Spectrum* (August 24, 1961), 192, 194, quotation on 194.

162. Carter interview.

163. Berman interview, 56.

164. McConnell to Ball, October 23, 1963. Also see "Summary Outline of US Frequency Allocation Proposals and Limitations for Space Users," October 1, 1962, COMSAT—Miscellaneous folder, box 7, Directors COMSAT Records 1962–66, Records of the Office of Emergency Planning, RG 396, NARA.

165. "Summary Outline of U.S. Frequency Allocation Proposals."

166. Telegram, Geneva to Secretary of State, October 22, 1963, TEL 6-1, Space Communication Frequencies, 10/1/63, ITU, folder, box 3654, Central Foreign Policy File, 1963, General Records of the Department of State, RG 59, NARA.

167. "Classified Report of the United States Territories Delegation to the EARC," 2.

168. Telegram, Geneva to Department of State, November 4, 1963, TEL 6-1, Space Communication Frequencies, 10/1/63, ITU, folder, box 3654, Central Foreign Policy File, 1963, General Records of the Department of State, RG 59, NARA.

169. C. W. Loeber to Thomas E. Nelson, October 17, 1963, Telecommunications: TEL 8-1, Radio Frequencies, 8/1/63, ITU, folder, box 3655, Central Foreign Policy File 1963, General Records of the Department of State, RG 59, NARA.

170. Telegram, Geneva to Department of State, November 4, 1963.

171. "Classified Report of the United States Territories Delegation to the EARC," 6.

172. Quoted in "Report on Geneva Space Radio Communication Conference," 186.

173. Quoted in Jonathan F. Galloway, *The Politics and Technology of Satellite Communications* (Lexington, Mass.: Lexington Books, 1972), 76.

174. Telegram, Geneva to Department of State, October 9, 1963, TEL 6-1, Space Communication Frequencies, 10/1/63, ITU, folder, box 3654, Central Foreign Policy File, 1963, General Records of the Department of State, RG 59, NARA.

175. Telegram, Geneva to Department of State, November 5, 1963, TEL 6-1, Space Communication Frequencies, 10/1/63, ITU, folder, box 3654, Central Foreign Policy File, 1963, General Records of the Department of State, RG 59, NARA.

176. "Report on Geneva Space Radio Communication Conference," 174, 186.

177. "Report on Geneva Space Radio Communication Conference," 174, 186.

178. Dean to Raish, October 25, 1963.

179. The overall authorization for satellite communications differed slightly in different regions of the globe. Of the three regions officially identified by the ITU, the block of 2800 MHz was for one of the regions in the Eastern Hemisphere; the other region in the East would have 2675 MHz for satellite communications service; the separate region of the Western Hemisphere had an authorization of 2600 MHz. See "Report on Geneva Space Radio Communication Conference," 178–79, 183.

180. "Report on Geneva Space Radio Communication Conference," 183.

181. "Report on Geneva Space Radio Communication Conference," 187.

182. "Report on Geneva Space Radio Communication Conference," 184.

183. On financial pressures on the United States as a motivation, see discussion in document I-41 in Logsdon, *Exploring the Unknown*, 2:165. On Soviet financial problems as motivation, see "Memorandum from the Deputy Director [of Intelligence], Central Intelligence Agency [Cline] to the President's Special Assistant for National Security Affairs [Bundy]," October 29, 1963, box 308, General, 10/63–11/63, Space Activities, Departments and Agencies Series, National Security Files, John F. Kennedy Library, reprinted as document 407 in U.S. Department of State, *Foreign Relations of the United States, 1961–63*, vol. 25.

184. "Memorandum Prepared in the Central Intelligence Agency," July 31, 1963, box 308, U.S.-USSR Cooperation 1961–63, Space Activities, Departments and Agencies Series, National Security Files, John F. Kennedy Library, reprinted as document 401 in U.S. Department of State, *Foreign Relations of the United States, 1961–63*, vol. 25.

185. See especially documents I-41 through I-43 in Logsdon, *Exploring the Unknown*, 2:165–82.

186. For an excellent overview of cooperative space activities between the United States and the Soviet Union during this period, see John M. Logsdon, "The Development of International Space Cooperation," in *Exploring the Unknown*, 2:11–13.

187. On November 15, Robert Kennedy anticipated another meeting between Khrushchev and the US president. The Soviets also anticipated better relations, had Kennedy lived. See Taubman, *Khrushchev*, 604.

188. "Report on Geneva Space Radio Communication Conference," 183–84, 186–87.

189. "Report on Geneva Space Radio Communication Conference," 183–84, 186–87. For official ITU report, see *Final Acts of the Extraordinary Administrative Radio Conference to Allocate Frequency Bands for Space Radiocommunication Purposes: Geneva, 1963* (Geneva: International Telecommunication Union, 1963), http://search.itu.int/history/HistoryDigitalCollectionDoc Library/4.89.43.en.100.pdf.

190. "Classified Report of the United States Territories Delegation to the EARC," 6.

191. "Report on Geneva Space Radio Communication Conference," 178. For 1980 predic-

tions, see "Proposals of the United States of America for the Extraordinary Administrative Radio Conference for Space Radiocommunication and Radio Astronomy (Geneva, 1963)," June 1, 1963, 13–14, box 821, docket 13522, docket files, Records of the FCC, NARA. For quotations, see "Additional Position Papers for the United States Delegation to the Extraordinary Administrative Radio Conference."

Chapter 4 · *Organizing the First Global Satellite Communications System*

1. "Agreement Establishing Interim Arrangements for a Global Commercial Communications Satellite System, Done at Washington August 20, 1964," Treaties and Other International Acts Series, Communications Satellite System (Comsat), Department of State (Washington, D.C.: Government Printing Office), located in folder 4, box 24, Joseph V. Charyk Papers, Special Collections Research Center, George Washington University, Washington, D.C.

2. Interview with William Gilbert Carter by Nina Gilden Seavey, July 15, 1985, 11–12, Comsat History Project, Series III: Comsat History Files, Comsat Corporation Collection, Milton S. Eisenhower Library Special Collections, Johns Hopkins University, Baltimore, Md.

3. UN General Assembly resolution 1721, section D.

4. The conference also recommended that Britain and Australia arrange talks with Europe before Commonwealth governments held further discussions. See *Commonwealth Conference on Satellite Communications, 1962*, vol. 1, *Report to the Governments*, sections A1, B1, B2, 13 April 1962, FO 371/17104, National Archives of the UK.

5. On activities of the State Department, see Senate Committee on Commerce, *Communications Satellite Incorporators: Hearings before the Committee on Commerce*, 88th Cong., 1st sess. (March 11, 1963), 24.

6. Carter interview, 15–16.

7. "Meeting Held at 10:00 A.M. September 18, 1962, in the Department of State to Consider British and Canadian Requests for Preliminary Talks on Communication Satellite Arrangements," September 20, 1962, 3–4, Meeker in the Department of State quoted on 4, Comsat—British/Canadian/U.S. Discussion folder, box 8, Directors COMSAT Records, 1962–66, Records of the Office of Emergency Planning, Record Group (RG) 396, National Archives and Records Administration (NARA), College Park, Md.

8. "Position Paper for the United States Delegation for the Meeting with Members of the European Conference on Satellite Communications and the Canadians, Rome—February 10, 1964," 17, TEL 6, Space Communications, 1/1/64, folder, box 1458, Central Foreign Policy File, 1964–66, General Records of the Department of State, RG 59, NARA. On developing countries increasingly playing an active role at the ITU during this period, see Hugh Richard Slotten, "The International Telecommunication Union, Space Radio Communications, and US Cold War Diplomacy, 1957–1963," *Diplomatic History* 37 (2013): 313–71. For a general history of UN governance during this period, including the broader debates, see Paul M. Kennedy, *The Parliament of Man: The Past, Present, and Future of the United Nations* (New York: Random House, 2006), and other sources included in the UN History Project, accessed April 6, 2021, https://www.histecon.magd.cam.ac.uk/unhist/research/bibliographies/governance.html.

9. "Summary Record of Preliminary Talks on Commercial Communications Satellites Held at the Department of State, October 29–31, 1962 among the United Kingdom, Canada and the United States of America," Frutkin quoted on 52–53, folder HOOO537, box 18, Comsat Corporation Collection, Milton S. Eisenhower Library Special Collections, Johns Hopkins University, Baltimore, Md.

10. "Meeting Held at 10:00 A.M. September 18, 1962, in the Department of State," 4.

11. W. G. Carter, "Notes for a Position Paper," October 6, 1962, 1–2, 1962 Satellite Act quoted on 1, Comsat—Miscellaneous folder, box 7, Directors COMSAT Records, 1962–66, Records of the Office of Emergency Planning, RG 396, NARA.

12. For first quotation see Blumenthal comments in "Summary Record of Preliminary Talks on Commercial Communications Satellites," 8, 51, Gardner quoted on 46.

13. Hope-Jones's comments reported in airgram from the US embassy in London to the Department of State, March 5, 1963, TEL-Telecommunications, UK, 2/1/63, folder, box 3660, Central Foreign Policy File 1963, General Records of the Department of State, RG 59, NARA.

14. This was the US attitude during at least one ITU meeting deciding about the international use of frequencies in this period. See Slotten, "International Telecommunication Union," 313–71. On ITU radio frequency conferences, see James Schwoch, *The American Radio Industry and Its Latin American Activities, 1900–1939* (Urbana: University of Illinois Press, 1990), 56–95; and James Schwoch, *Global TV: New Media and the Cold War, 1946–69* (Urbana: University of Illinois Press, 2009), 18–25, 31–35.

15. On the historical context to European interest in space, see especially Lorenza Sebesta, "US-European Relations and the Decision to Build Ariane, the European Launch Vehicle," in *Beyond the Ionosphere: Fifty Years of Satellite Communication*, ed. Andrew J. Butrica (Washington, D.C.: NASA History Office, 1997), 137–39. Also see the European Space Agency's History Study Reports, including especially John Krige and Arturo Russo, *Reflections on Europe in Space* (Noordwijk, Netherlands: ESA, 1994). On the importance of the technology gap for US-European relations, see Lorenza Sebesta, *United States-European Cooperation in Space during the Sixties* (Noordwijk, Netherlands: ESA, 1994), 18, 20–21.

16. David Price, British Member of Parliament, quoted by Michelangelo De Maria, *The History of ELDO Part 1: 1961–1964* (Noordwijk, Netherlands: ESA, 1993), 16.

17. John Krige, *The Launch of ELDO* (Noordwijk, Netherlands: ESA, 1993), 11,15.

18. On European interest in space and fears of a technology gap with the United States, see John Krige, Angelina Long Callahan, and Ashok Maharj, *NASA in the World: Fifty Years of International Collaboration in Space* (New York: Palgrave Macmillan, 2013), 52–53. For Europe and beginnings of space program, see Sebesta, "U.S.-European Relations and the Decision to Build Ariane," 137–39; John Krige and Arturo Russo, *A History of the European Space Agency, 1958–1987*, vol. 1, *The Story of ESRO and ELDO—1958–1973* (Noordwijk, Netherlands: ESA, 2000); and John Krige, "Building Space Capability through European Regional Collaboration," in *Remembering the Space Age: Proceedings of the 50th Anniversary Conference*, ed. Steven J. Dick (Washington, D.C.: NASA, 2008), 37–52.

19. For discussion of early international activities of NASA, see Arnold W. Frutkin, *International Cooperation in Space* (Englewood Cliffs, N J : Prentice-Hall, 1965).

20. Lorenza Sebesta, "U.S.-European Relations and the Decision to Build Ariane," 138, 142–43. On the British Commonwealth system, see especially Nigel Wright, "The Formulation of British and European Policy toward an International Satellite Telecommunications System," in Butrica, *Beyond the Ionosphere*, 157–69. For discussion of example of two locations in India, see memorandum, Joseph V. Charyk, April 1, 1964, TEL 6, Space Communications, 4/1/64, folder, box 1458, Central Foreign Policy File, 1964–66, General Records of the Department of State, RG 59, NARA.

21. Wright, "Formulation of British and European Policy," 157–60.

22. Ronald C. Hope-Jones to R. H. Oakeley, June 13, 1963, AVIA 92/162, National Archives of the UK.

23. See D. J. Gibson, "Commonwealth Conference on Communications Satellites, February 1962: Preparation of U.K. Policy," December 15, 1961, FO 371/165275, National Archives of the UK.

24. Wright, "Formulation of British and European Policy," 160–63.

25. Wright, "Formulation of British and European Policy," 160–63.

26. J. Mark, "Satellite Communications," March 7, 1962, T 319/157, National Archives of the UK. Also see Wright, "Formulation of British and European Policy," 161–63. For background on coordination of Commonwealth communication during the 1950s, see Simon James Potter,

Broadcasting Empire: The BBC and the British World, 1922–1970 (Oxford: Oxford University Press, 2012), 175–99.

27. R. F. G. Sarell to "Mr. Melville," March 19, 1962, FO 371/165275, National Archives of the UK.

28. Wright, "Formulation of British and European Policy," 163–65; Hope-Jones quotation on 164n29.

29. Ronald C. Hope-Jones, "Commonwealth Conference on Satellite Communications," March 15, 1962, FO 371/165275, National Archives of the UK. Also see Wright, "Formulation of British and European Policy," 163–65.

30. Ronald C. Hope-Jones, "Commonwealth Conference on Satellite Communications."

31. On activities of State Department, see Senate Committee on Commerce, *Communications Satellite Incorporators*, 24.

32. Krige and Russo, *History of the European Space Agency*, vol. 1, chapters 1 and 3.

33. Krige and Russo, *History of the European Space Agency*, vol. 1, Amaldi quotation ("mere spectators") on 20, last quotations on 91.

34. Krige and Russo, *History of the European Space Agency*, 1:28, 30.

35. Comments of CNET director, related by Edgar L. Piret to Secretary of State, March 24, 1963, TEL 6, 3/1/73 [sic], folder, box 3654, Central Foreign Policy File, 1963, General Records of the Department of State, RG 59, NARA. For further information about the ministerial meeting, see airgram, Edgar L. Piret to secretary of state, March 28, 1963, SP—Space and Astronautics, 2/1/63, FALK IS, folder, box 4185, Central Foreign Policy File 1963, General Records of the Department of State, RG 59, NARA.

36. Sebesta, "U.S.-European Relations and the Decision to Build Ariane," 138, 142–43.

37. For quotation, see M. W. Hodges to James F. Hosie, "Post Office Working Party on Satellite Communications," December 10, 1962, CAB 124/2281, National Archives of the UK. On France not yet having a definite policy, see P. Dixon, Paris, to Foreign Office, November 13, 1962, FO 371/165249, National Archives of the UK.

38. M. D. Butler to Ronald C. Hope-Jones, December 3, 1962, PO 371,165249, National Archives of the UK.

39. Dixon, Paris, to Foreign Office, November 13, 1962.

40. Butler to Hope-Jones, December 3, 1962.

41. James F. Hosie to Roger Nathaniel Quirk, "Satellite Communications," December 11, 1962, CAB 124/2281, National Archives of the UK.

42. M. W. Hodges to James F. Hosie, "European Satellite Communications Organization?," December 6, 1962, CAB 124/2281, National Archives of the UK.

43. Wright, "Formulation of British and European Policy," 165–69. On the composition of the ad hoc committee, see C. W. Loeber to C. E. Lovell, February 6, 1963, TEL 6, Space Communications, 2/1/63, folder, box 3654, Central Foreign Policy File, 1963, General Records of the Department of State, RG 59, NARA.

44. For view of UK Treasury official, see A. M. Bailey, "Civil Satellite Communications," December 11, 1962, T 225/2449, National Archives of the UK.

45. Ronald C. Hope-Jones to G. A. Garey-Foster, January 23, 1963, FO 371/165249, National Archives of the UK.

46. Mathieu Segers, "De Gaulle's Race to the Bottom: The Netherlands, France and the Interwoven Problems of British EEC Membership and European Political Union, 1958–1963," *Contemporary European History* 19 (2010): 119–20.

47. "British Discussions with the Conference of European Postal and Telecommunications Administrations (CEPT) Cologne, December 1962," FO 371/17/046, National Archives of the UK.

48. Ronald C. Hope-Jones, "Satellite Communications," December 21, 1962, FO 371/171046,

National Archives of the UK. Also see Wright, "Formulation of British and European Policy," 165–69.

49. Edward A. Bolster to John S. Meadows, February 5, 1963, TEL-Telecommunications, UK, 2/1/63, folder, box 3660, Central Foreign Policy File, 1963, General Records of the Department of State, RG 59, NARA.

50. Airgram, US embassy in London to the Department of State, March 5, 1963, TEL-Telecommunications, UK, 2/1/63, folder, box 3660, Central Foreign Policy File, 1963, General Records of the Department of State, RG 59, NARA.

51. Hope-Jones's comments reported in airgram from the US embassy in London to the Department of State, March 5, 1963.

52. Views of Canadian mission in London reported in memorandum, "British Policy Re Commercial Communications Satellites and Canadian Attitude Thereto," March 29, 1963, Telecommunications: TEL 3, Organizations and Conferences, 2/1/63, ITU, folder, box 3660, Central Foreign Policy File, 1963, General Records of the Department of State, RG 59, NARA.

53. R. C. Hope-Jones to P. Reilly, "The Canadians and Satellite Communications," March 26, 1963, FO 371/171048, National Archives of the UK.

54. Airgram, Thomas E. Nelson to secretary of state, March 26, 1963, TEL 6, 3/1/73 [*sic*], folder, box 3660, Central Foreign Policy 1963, General Records of the Department of State, RG 59, NARA.

55. Telegram, Paris to secretary of state, March 13, 1963, TEL-Telecommunications, UK, 2/1/63, folder, box 3660, Central Foreign Policy File, 1963, General Records of the Department of State, RG 59, NARA.

56. The scientific attaché at the US embassy in Paris reported to the State Department about the Paris meeting based on the comments of a French participant, Marzin, the director of the CNET. See airgram, Piret to secretary of state, March 24, 1963. The final examination of the queries before submittal to the United States was made at a meeting of the ad hoc committee of the Telecommunications Commission of the CEPT in Karlsruhe, Germany (June 5–7, 1963). For minutes of this meeting, see enclosures 1 and 2 included in communication from Bonn embassy to Department of State, June 25, 1963, Telecommunications: TEL 3, Organization and Conferences CEPT folder, box 3652, Central Foreign Policy File, 1963, General Records of the Department of State, RG 59, NARA.

57. On Graham, see statement based on interview in Jonathan F. Galloway, *The Politics and Technology of Satellite Communications* (Lexington, Mass.: Lexington Books, 1972), 81.

58. US embassy in London to secretary of state, May 16, 1963, Telecommunications: TEL 3, Organizations and Conferences, CEPT folder, box 3652, Central Foreign Policy File, 1963, General Records of the Department of State, RG 59, NARA.

59. The views of Welch and the secretary of state are related in memorandum, "Plans of the Communications Satellite Corporation to Establish Contacts with the Operating Communications Agencies of Canada and the Principal European Countries," May 17, 1963, TEL-Telecommunications, CAN-A, folder, box 3657, Central Foreign Policy File, 1963, General Records of the Department of State, RG 59, NARA.

60. Memorandum, "Report on Conversations by Messrs. Welch and Charyk of the Communications Satellite Corporation with officials of the Canadian Department of Transport," May 27, 1963, TEL-Telecommunications, CAN-A, folder, box 3657, Central Foreign Policy File, 1963, General Records of the Department of State, RG 59, NARA.

61. Memorandum, Philip H. Trezise to secretary of state, May 15, 1963, TEL 6, 3/1/73 [*sic*], folder, box 3654, Central Foreign Policy File, 1963, General Records of the Department of State, RG 59, NARA.

62. Claim by FCC official related in memorandum, "Developments Subsequent to the Secretary's Meeting on May 17 with Messrs. Welch and Charyk of the Communications Satellite

Corporation," May 24, 1963, TEL 6, 3/1/73 [*sic*], folder, box 3654, Central Foreign Policy File, 1963, General Records of the Department of State, RG 59, NARA. For another example of a US government official critical of the State Department for not taking an active role in Comsat foreign negotiations, see comments of Senator Fulbright recounted in E. C. Welsh to George Ball, October 2, 1963, TEL 6, 6/1/63, folder, box 3654, Central Foreign Policy File, 1963, General Records of the Department of State, RG 59, NARA.

63. On discussion of equipment problems and four categories, which were clearly flexible, see airgram, Paris embassy to Department of State, June 2, 1963, "Global Communication Satellites System: Meeting of Welch and Charyk with CNET Officials," TEL 6, 6/1/63, folder, box 3654, Central Foreign Policy File, 1963, General Records of the Department of State, RG 59, NARA; and airgram, Rome embassy to Department of State, June 5, 1963, "Communications Satellite System," TEL-Telecommunications 2/1/63 US folder, box 3660, Central Foreign Policy File, 1963, General Records of the Department of State, RG 59, NARA.

64. Quotation from report of meeting with Comsat executives in enclosure 2 (annex 1 to minutes of the ad hoc committee meeting, June 5–7, 1963) included in communication from Bonn embassy to Department of State, June 25, 1963.

65. Airgram, Rome embassy to Department of State, June 5, 1963; airgram, Paris embassy to Department of State, June 2, 1963.

66. Airgram, Paris embassy to Department of State, June 2, 1963.

67. Airgram, Paris embassy to Department of State, June 2, 1963.

68. Airgram, Paris embassy to Department of State, June 2, 1963.

69. Carter interview, 57.

70. Carter interview, 22.

71. Airgram, Paris embassy to Department of State, June 2, 1963.

72. "A Progress Report on Planning for a Global Commercial Communications Satellite System," n.d. [May 1964], 2, 10–11, 14, Comsat International Agreements folder, box 8, Directors COMSAT Records, 1962–66, Records of the Office of Emergency Planning, RG 396, NARA.

73. Aide-Memoire, June 26, 1963, attached as Annex B with memorandum, Department of State to Circular, "Commercial Communications Satellite System," June 27, 1963, TEL 6, Space Communications, 2/1/63, folder, box 3654, Central Foreign Policy File, 1963, General Records of the Department of State, RG 59, NARA.

74. On organization of Ad Hoc Communications Satellite Group, see "Report of Director of Telecommunications Management," August 26, 1963, in House Committee on Interstate and Foreign Commerce, *Report No. 809, Communications Satellite Act of 1962: The First Year*, 88th Cong., 1st sess. (October 3, 1963), 29.

75. Memorandum to the Ad Hoc Communications Satellite Group, "Redraft of Communications Satellite Corporation Principles for Establishment of Global Commercial Communications Satellite System," August 13, 1963, TEL 6, Space Communications, 2/1/63, folder, box 3654, Central Foreign Policy File, 1963, General Records of the Department of State, RG 59, NARA. On Europeans favoring multilateral agreements and opposing bilateral arrangements, see Warren B. Cheston to Department of State, August 20, 1963, SP—Space and Astronautics, 2/1/63, FALK IS, folder, box 4185, Central Foreign Policy File, 1963, General Records of the Department of State, RG 59, NARA.

76. Italian account of comments related in telegram, US embassy in Rome to secretary of state, May 6, 1963, Telecommunications: TEL 3, Organizations and Conferences, CEPT, folder, box 3652, Central Foreign Policy File, 1963, General Records of the Department of State, RG 59, NARA.

77. A. Rumbold to Patrick Reilly, May 22, 1963, FO 371/171056, National Archives of the UK.

78. Ronald C. Hope-Jones to J. A. Robinson, September 17, 1963, FO 371/171061, National Archives of the UK.

79. US embassy in London to Department of State, October 15, 1963, TEL—Telecommunications, F, 2/1/63, folder, box 3658, Central Foreign Policy File, 1963, General Records of the Department of State, RG 59, NARA.

80. "Satellite Communication Conference Report by the United Kingdom Delegation," July 26, 1963, CAB 130/191, National Archives of the UK.

81. Hope-Jones to Robinson, September 17, 1963.

82. "Satellite Communication Conference Report by the United Kingdom Delegation."

83. Hope-Jones to Robinson, September 17, 1963.

84. S. Meijer to chairman of the London Session of the European Conference on Telecommunication by Satellites, "Summary of letter of transmittal," October 21, 1963, included with US mission in Geneva to Department of State, "Communication Satellites: Summary of Luncheon meeting," November 1, 1963, TEL 6-1, Space Communication Frequencies, 10/1/63, ITU, folder, box 3654, Central Foreign Policy File, 1963, General Records of the Department of State, RG 59, NARA. On proposal to have Comsat manage system under guidance of administrative council, see US embassy in London to secretary of state, October 15, 1963, TEL 6-1, Space Communication Frequencies, 10/1/63, ITU, folder, box 3654, Central Foreign Policy File, 1963, General Records of the Department of State, RG 59, NARA.

85. Views of Patrick Reilly related in US embassy in London to secretary of state, October 15, 1963.

86. Hope-Jones comments retold in airgram, John S. Meadows to Department of State, June 1, 1963, TEL 6, 6/1/63, folder, box 3654, Central Foreign Policy File, 1963, General Records of the Department of State, RG 59, NARA.

87. Airgram, Paris embassy to Department of State, June 2, 1963.

88. Airgram, Rome embassy to Department of State, June 5, 1963.

89. US embassy in Paris to Department of State, "Global Communication Satellite System: Second Meeting of the European Conference on Satellite Communications," August 9, 1963, SP—Space and Astronautics, 2/1/63, FALK IS, folder, box 4185, Central Foreign Policy File, 1963, General Records of the Department of State, RG 59, NARA.

90. "Towards a European Organization for Telecommunications through Satellites: Work of the Conference in Rome," November 30, 1963, included as enclosure 3 with US embassy in Rome to Department of State, December 18, 1963, Telecommunications: TEL 3, Organizations and Conferences, 2/1/63, folder, box 3652, Central Foreign Policy File, 1963, General Records of the Department of State, RG 59, NARA.

91. D. P. Reilly, "European Satellite Communications Conference," November 14, 1963, FO 371/171066, National Archives of the UK.

92. "Towards a European Organization for Telecommunications through Satellites." On decision about the committee, see "Summary of European Regional Organization in the Communications Satellite Field," December 9, 1963, 4, included with Egidio Ortona to Dean Rusk, November 30, 1963, TEL 6, 6/1/63, folder, box 3654, Central Foreign Policy File, 1963, General Records of the Department of State, RG 59, NARA. For quotation, see memorandum, E. G. Griffith Johnson to George W. Ball, December 4, 1963, TEL 6, 6/1/63, folder, box 3654, Central Foreign Policy File, 1963, General Records of the Department of State, RG 59, NARA.

93. Airgram, Department of State to European embassies, December 13, 1963, TEL 6, 6-1, folder, box 3654, Central Foreign Policy File, 1963, General Records of the Department of State, RG 59, NARA.

94. Airgram, US embassy in Rome to Department of State, December 18, 1963.

95. J. E. Dingman to Leo Welch, December 6, 1963, TEL 6, 6/1/63, folder, box 3654, Central Foreign Policy File, 1963, General Records of the Department of State, RG 59, NARA. As late as the end of October, State Department officials in Europe still warned of a "dissident U.K. faction" within the British aerospace industry, centered on an organization known as the British

Space Development Corporation, which was interested in building a second, independent satellite system. See airgram, Edgar L. Piret to Department of State, October 24, 1963, Telecommunications: TEL 3, Organizations and Conferences, 2/1/63, folder, box 3652, Central Foreign Policy File, 1963, General Records of the Department of State, RG 59, NARA.

96. Memorandum, Ralph Clark to Ralph Dungan, "International Telephone Cables and Communications Satellites," November 14, 1963, FG 11–6–2 Office of the Director of Telecommunications Management folder, box 115, White House Central Subject File, John F. Kennedy Presidential Library, Boston, Mass.

97. Memorandum, Johnson to Ball, December 4, 1963.

98. Dingman to Welch, December 6, 1963. For the State Department belief that the Europeans saw this as an "either-or-choice," see memorandum, Abram Chayes and G. Griffith Johnson to acting secretary of state, "Your Meeting with Mr. Kappel and Dr. Wiesner," December 13, 1963, TEL 6, 6/1/63, folder, box 3654, Central Foreign Policy File, 1963, General Records of the Department of State, RG 59, NARA.

99. Memorandum, Chayes and Johnson to Acting Secretary of State, December 13, 1963.

100. Memorandum of conversation, Department of State, "Meeting of December 16, 1963 with Chairman of the Board of the American Telephone & Telegraph Company to discuss Commercial Communication Satellite Program," December 16, 1963, TEL—Telecommunications, 2/1/63 US, folder, box 3660, Central Foreign Policy File, 1963, General Records of the Department of State, RG 59, NARA.

101. Memorandum, Chayes and Johnson to Acting Secretary of State, December 13, 1963.

102. Memorandum of conversation, Department of State, "Meeting of December 16, 1963."

103. The British apparently began to reassess their position in support of cables even before talks in Europe with AT&T. On this issue and on bilateral talks AT&T conducted in Europe, see airgram, John S. Meadows to Department of State, December 21, 1963, Telecommunications: TEL 3, Organization and Conferences, CEPT, folder, box 3652, Central Foreign Policy File, 1963, General Records of the Department of State, RG 59, NARA.

104. Memorandum of conversation, Department of State, "Meeting of December 16, 1963"; Johnson to Ball, December 4, 1963; Ronald C. Hope-Jones to M. D. Butler, January 23, 1964, FO 371/176286, National Archives of the UK.

105. Harlan Cleveland to Abram Chayes, February 7, 1964, TEL 6, Space Communications, 1/1/64, folder, box 1458, Central Foreign Policy File, 1964–66, General Records of the Department of State, RG 59, NARA.

106. Abram Chayes to undersecretary of the Department of State, February 4, 1964, TEL 6, Space Communications, 1/1/64, folder, box 1458, Central Foreign Policy File, 1964–66, General Records of the Department of State, RG 59, NARA.

107. "Position Paper for the United States Delegation for the Meeting with Members of the European Conference," 4–5, 12–13.

108. "Position Paper for the United States Delegation for the Meeting with Members of the European Conference," 17.

109. Airgram, Meadows to Department of State, December 21, 1963.

110. Airgram, Meadows to Department of State, December 21, 1963, 6, 9, 14, 17.

111. On how these countries contributed to the success of the negotiations with the Europeans, see interview with John A. Johnson by Nina Gilden Seavey, January 12, 1983, 7–8, Comsat History Project, Series III: Comsat History Files, Comsat Corporation Collection, Milton S. Eisenhower Library Special Collections, Johns Hopkins University, Baltimore, Md. On Canada deciding to support US efforts, see cabinet minutes, no. 42–63, August 3, 1963, 22, RG 2, A-5-a, box 6254, Archives and Library Canada, Ottawa, Ontario.

112. Document titled "Introduction" attached to [Agency for International Development] AID/Washington to embassies, "Space Communications," October 22, 1964, Tel 6, Space Com-

munications, 10/1/64, folder, box 1457, Central Foreign Policy File, 1964–66, General Records of the Department of State, RG 59, NARA.

113. "Introduction" attached to AID/Washington to embassies, "Space Communications."

114. "Introduction" attached to AID/Washington to embassies, "Space Communications."

115. "Introduction" attached to AID/Washington to embassies, "Space Communications."

116. Western European countries also benefited more than most other countries in the world from a weighted system of voting. The UK explicitly looked to the IMF as a model for the global satellite communications system. See Patrick Reilly comments, meeting minutes, Cabinet Ministerial Committee on Satellite Communications, November 22, 1963, 1, CAB 134/1566, National Archives of the UK.

117. "Position Paper for the United States Delegation for the Meeting with Members of the European Conference on Satellite Communication, Canada, Japan, and Australia, London— June 12–13 and 18–20, 1964," n.d., 2–3, "Tel 6, Space Communications, 6/1/64, folder 1 of 2, box 1458, Central Foreign Policy File, 1964–66, General Records of the Department of State, RG 59, NARA. John A. Johnson interview, 11.

118. "Introduction" attached to AID/Washington to embassies, "Space Communications." Nations choosing not to join could still lease communication channels from the system.

119. Washington, D.C. to UK Foreign Office, "Satellite Communications—Voting Procedure," July 23, 1964, FO 371/176291, National Archives of the UK.

120. Charles Johnston comments, meeting minutes, Cabinet Ministerial Committee on Satellite Communications, June 22, 1964, 4, CAB 134/1566, National Archives of the UK.

121. Provisional summary record—second plenary session, July 22, 1964, 2–3, Plenipotentiary Conference to Establish Interim Arrangements for a Global Commercial Communications Satellite System, Comsat International Agreement folder, box 8, Directors COMSAT Records, 1962–66, Records of the Office of Emergency Planning, RG 396, NARA. For British perspective on the European position, see Ronald C. Hope-Jones to R. S. Faber, July 8, 1964, FO 371/176291, National Archives of the UK.

122. Annex V, "Agreement Establishing Interim Arrangements."

123. Lee R. Marks to James D. O'Connell, July 29, 1964, Defense-Comsat Corp. Negotiations folder, box 8, Directors COMSAT Records, 1962–66, Records of the Office of Emergency Planning, RG 396, NARA.

124. Reilly comments, meeting minutes, Cabinet Ministerial Committee on Satellite Communications, November 22, 1963, 1.

125. Marks to O'Connell, July 29, 1964.

126. On the development of technological infrastructure in Europe, see, for example, Erik van der Vleuten and Arne Kaijser, eds., *Networking Europe: Transnational Infrastructures and the Shaping of Europe, 1850–2000* (Sagamore Beach, Mass.: Science History, 2006); Fickers and Griset, *Communicating Europe*; Balbi and Fickers, *History of the International Telecommunication Union*; and all the volumes in the Making Europe series published by Palgrave Macmillan.

127. The United States relied on sympathetic Canadian embassy officials in London for information about internal British political developments. See "British Policy Re Commercial Communications Satellites and Canadian Attitude Thereto."

Conclusion

1. International agreements allocating radio frequencies for communications satellites continued to be decided at special meetings of the ITU, but the needs of the Intelsat global system heavily influenced the United States' involvement and ITU's decisions. See Hugh Richard Slotten, "The International Telecommunication Union, Space Radio Communications, and U.S. Cold War Diplomacy, 1957–1963," *Diplomatic History* 37 (2013): 313–71.

2. John Downing, "The Intersputnik System and Soviet Television," *Soviet Studies* 37 (October 1985): 465–83.

3. National Security Action Memorandum No. 342, March 4, 1966, Tel 6, 4/1/66, folder, box 1455, Central Foreign Policy File, 1964–66, Records of the Department of State, Record Group (RG) 59, National Archives and Records Administration (NARA), College Park, Md.

4. "Progress in the Field of Commercial Communication Satellites—Intelsat," March 29, 1966, Tel 6, 4/1/66, folder, box 1455, Central Foreign Policy File, 1964–66, Records of the Department of State, RG 59, NARA.

5. See memorandum, Bureau of African Affairs, Dept. of State, David D. Newsom to Samuel Z. Westerfield, September 8, 1964, Tel 6, 8/2/64, folder, box 1457, Central Foreign Policy File 1964–66, General Records of the Department of State, RG 59, NARA. Also see Joseph N. Pelton, *Global Communications Satellite Policy: Intelsat, Politics and Functionalism* (Mt. Airy, Md.: Lomond Books, 1974), 131.

6. "Kenya Enters the Space Age: President Talks to U Thant by Satellite," *East African Standard*, November 13, 1970.

7. James D. O'Connell, "A Global System of Satellite Communications—The Hazards Ahead," n.d. (circa February 1967), UT1: Communications—Telecommunications folder, box 96, White House Central Files—Confidential File, Lyndon B. Johnson Presidential Library, Austin, Texas.

8. Sara Fletcher Luther, *The United States and the Direct Broadcast Satellite: The Politics of International Broadcasting in Space* (New York: Oxford University Press, 1988); Edward A. Comor, *Communication, Commerce, and Power: The Political Economy of America and the Direct Broadcast Satellite, 1960–2000* (New York: St. Martin's Press, 1998). On the SITE experiment, see, for example, Daya Kishan Thussu, *International Communication: Continuity and Change*, 2nd ed. (London: Hodder Arnold, 2006), 27–31.

9. Daya Kishan Thussu, "Reinventing 'Many Voices': MacBride and a Digital New World Information and Communication Order," *Journal of European Institute for Communication and Culture* 22 (2015): 252–63, quotation on 255.

10. For television, see especially two early UNESCO studies: Kaarle Nordenstreng and Tapio Varis, *Television Traffic—A One-Way Street? A Survey and Analysis of the International Flow of Television Programme Material* (Paris: Unesco, 1974); and Tapio Varis, *International Flow of Television Programmes* (Paris: Unesco, 1985).

11. Joseph D. Straubhaar, *World Television: From Global to Local* (Los Angeles: Sage, 2007), 262, table A-2. There is a need for more research on the connection between Intelsat and global television.

12. Juan Somavia quoted in Thussu, "Reinventing 'Many Voices,'" 252. On "media imperialism," see Kalyani Chadha and Anandam Kavoori, "The New Normal: From Media Imperialism to Market Liberalization—Asia's Shifting Television Landscapes," *Media, Culture and Society* 37 (2015): 479–92; and Oliver Boyd-Barrett, *Media Imperialism* (Los Angeles: Sage, 2015). A similar view by a communication studies scholar from this early period that mainly focuses on the imperial ambitions of the United States specifically emphasizes US dominance of Intelsat but does not clearly link this to media and communications content: Herbert I. Schiller, *Mass Communications and American Empire* (New York: Augustus M. Kelley, 1969), 127–46.

13. Quoted by Cees J. Hamelink, "Infomatics: Third World Call for New Order," *Journal of Communication* (September 1979): 144–48, on 145.

14. International Commission for the Study of Communication Problems, *Many Voices, One World: Towards a New, More Just, and More Efficient World Information and Communication Order* (Paris: Unesco, 1980), https://unesdoc.unesco.org/ark:/48223/pf0000040066.

15. UNESCO, *Records of the General Assembly*, 21st session, Belgrade, vol. 1 (Paris: Unesco, 1980), 68–71, https://unesdoc.unesco.org/ark:/48223/pf0000114029/PDF/114029eng.pdf.multi.

16. UNESCO, *Records of the General Assembly*, 21st session, 1:71.

17. Department of State airgram, "Market for Satellite Earth Station Equipment and Related Engineering Services," March 1, 1967, Tel 6, 3/1/67, folder, box 1457, Central Foreign Policy Files, 1967–69, Records of the Department of State, RG 59, NARA.

18. Department of State telegram, "Moroccan Request for USG Utilization of Moroccan Satellite Communications Facilities," February 7, 1970, box 18 of 23, INTELSAT Conference Documents and Files, 1968–71, Records of the Department of State, RG 59, NARA.

19. Department of State telegram, "Continued Moroccan Interest in USG Utilization of Moroccan Satellite Communications Facilities," May 25, 1970, Morocco folder, box 18 of 23, INTELSAT Conference Documents and Files, 1968–71, Records of the Department of State, RG 59, NARA.

20. Department of State telegram, "Moroccan Request for USG Utilization."

21. Quoted by Heather E. Hudson, *Global Connections: International Telecommunications Infrastructure and Policy* (New York: Wiley, 1997), 361.

22. Hudson, *Global Connections*, 361.

23. Hudson, *Global Connections*, 365–66.

24. Joseph N. Pelton, "Project SHARE and the Development of Global Satellite Communications," in *Beyond the Ionosphere: Fifty Years of Satellite Communication*, ed. Andrew J. Butrica (Washington, D.C.: NASA History Office, 1997), 257–64.

25. Heather E. Hudson, *Communication Satellites: Their Development and Impact* (New York: Free Press, 1990), 7, 63–64, 94, 185, 190, 192, 204, 209.

26. Mohammad Razani, *Commercial Space Technologies and Applications: Communication, Remote Sensing, GPS, and Meteorological Satellites*, 2nd ed. (Boca Raton, Fla.: CRC Press, 2018), 182.

27. "Multilateral Agreement relating to the International Telecommunications Satellite Organization INTELSAT (with annexes)," August 20, 1971, https://treaties.un.org/doc/Publication/UNTS/Volume%201220/volume-1220-I-19677-English.pdf.

28. Joseph N. Pelton, "History of Satellite Communications," in *Handbook of Satellite Applications*, ed. Joseph N. Pelton, Scott Madry, and Sergio Camacho-Lara (Cham, Switzerland: Springer, 2017), 31–71, on 47, 51–52.

29. Hudson, *Communication Satellites*, 188, 199; Hudson, *Global Connections*, 259, 388.

30. Thussu, *International Communication*, 35–34, 66–75.

31. Pelton, "History of Satellite Communications," 52–53.

32. Quoted by Hudson, *Global Connections*, 381.

33. Hudson, *Global Connections*, 382–83; Hudson, *Communication Satellites*, 191–92.

34. Bruce R. Elbert, *Introduction to Satellite Communication*, 2nd ed. (Boston: Artech House, 1999), 54, 56.

35. Jeremiah Hayes, "A History of Transatlantic Cables," *IEEE Communications Magazine* 46 (September 2008): 12–18.

36. Hudson, *Global Connections*, 376–79; Nicole Starosielski, *The Undersea Network* (Durham: University of North Carolina Press, 2015), 47.

37. *Aeronautics and Space Report of the President: Fiscal Year 1994 Activities* (Washington, D.C.: NASA, 1995), 33.

38. Hudson, *Global Connections*, 368, 375; Jonathan Reed Winkler, "Bridging the Gap: The Cable and Its Challenges," in *Communications Under the Seas: The Evolving Cable Network and Its Implications*, ed. Bernard Finn and Daqing Yang (Cambridge, Mass.: MIT Press, 2009), 25–44, on 41.

39. Starosielski, *Undersea Network*, 53.

40. Jeff Hecht, "Fiber-Optic Submarine Cables: Covering the Ocean Floor with Glass," in Finn and Yang, *Communications Under the Seas*, 45–58, on 53. Also see Pelton, "History of Satellite Communications," 67–68.

41. Hudson, *Global Connections*, 368–69.

42. Winkler, "Bridging the Gap," 43.

43. Hudson, *Global Connections*, 369.

44. Robert J. Oslund, "The Geopolitics and Institutions of Satellite Communications," in *Communications Satellites: Global Change Agents*, ed. Joseph N. Pelton, Robert J. Oslund, and Peter Marshall (Mahwah, N.J.: Erlbaum, 2004), 111–45, on 140–41.

45. "About Us," ITSO, June 24, 2020, https://itso.int/about-us/. For official wording of Resolution 1721, see "Resolution Adopted by the General Assembly," 1721 (XVI), International Co-operation in the Peaceful Uses of Outer Space," UN Office for Outer Space Affairs, December 20, 1961, https://www.unoosa.org/oosa/en/ourwork/spacelaw/treaties/resolutions/res_16_1721.html.

46. Martin Collins, *A Telephone for the World: Iridium, Motorola, and the Making of a Global Age* (Baltimore, Md.: Johns Hopkins University Press, 2018), 2.

de la Grandville, Jean, 154
deregulation. *See* neoliberalism
developing countries. *See* Global South
development, 13, 19, 189; Intelsat's commitment to, 184–85; ITU and Global South and, 50; Lyndon Johnson's commitment to, 72, 180–82; role of communications in, 121, 123. *See also* assistance, technical; Global South; modernization theory
de Wolf, Francis Colt, 117, 120, 127–28, 134
Dingman, James E., 114
disarmament. *See* arms control
Doerfer, John C., 57, 76
Donner, Frederic G., 114
Dryden, Hugh L., 71, 87, 131
Dutton, Frederick, 76, 83–84, 88–89, 91

Early Bird. See *Intelsat I*
Earth station, communications satellite. *See* ground station, communications satellite
Eastern and Associated Telegraph Companies, 6–7
East-West conflict. *See* Cold War
Echo 1, 30–31, 34–36, 67, 69
Eisenhower, Dwight D., 16, 28–29, 34, 44; "total Cold War" and, 21, 28, 50–51
Eisenhower administration, 4, 95–6; communications satellite policy and, 22, 30, 33–40, 64; IRAC and, 44–45; peaceful uses of outer space and, 57; response to Sputnik, 28
Ende, Asher, 63–64
European Conference of Postal and Telecommunications Administrations (CEPT), 155, 185
European Conference on Satellite Communications (CETS): desire for multilateral talks with US, 171; pressured by AT&T to support US government proposals, 166–69; separate system idea, 163; wants Intelsat to support local manufacturing in Europe, 164, 174–75. *See also* Intelsat; regionalism
European Economic Community (EEC), 153, 155
European Space Agency, 185
European Space Research Organization (ESRO), 150, 153
European Space Vehicle Launcher Development Organization (ELDO), 150, 152–53, 156, 165
European Telecommunications Satellite Organization (Eutelsat), 185–86
EUROSPACE, 150

Federal Communications Commission (FCC): policy toward satellite communications after 1961, 73–85, 89–92, 96–101; policy toward satellite communications before 1961, 31, 33, 38; relationship to executive agencies, 62–63, 80; relationship with IRAC, 37–38; role in initial Intelsat negotiations, 147, 159, 169; role in ITU conferences, 40, 44–47, 115, 131; Separate Systems decision, 187
Fensterwald, Bernard, 97–99
fiber-optic cables, undersea, 187–88
filibuster, 93, 96–100
Ford Aerospace, 185
France: early communications satellite experiments with US, 2, 69; initial Intelsat negotiations and, 146, 149–50, 154–55, 160, 163–66, 172–75, 181; involvement in Intelsat during the 1990s, 184; involvement in work of ITU, 119, 122, 124, 137; involvement with cable and wireless before World War II, 7–8, 12
"free flow of information," 186
Frutkin, Arnold, 91, 148

Gagarin, Yuri Alekseyevich, 13, 53–54, 56, 58, 70
Gardner, Richard N., 117, 125, 149
Gardner Report, 62
Geneen, Harold S., 77
General Electric (GE), 9, 30, 66
General Telephone and Electronics (GTE), 33, 65–66, 68, 73–74, 119
Germany, 7–8, 12, 110; West, 124–25, 150, 160, 172, 184
Gilpatric, Roswell L., 82, 110
Glennan, T. Keith, 30, 33–37
Global South, 13–14, 106, 178–79; cultural imperialism and, 182–83; in initial Intelsat negotiations, 18, 145, 148, 175, 180; involvement in ITU, 50, 141
globalism, nationalist, 19, 179, 189
Globalstar, 189
Gore, Albert, 95–96, 98–99
Graham, Beardsley, 66, 68–69
Graham, Martha, 112
Graham, Philip, 109–10, 112, 158
Green, Donald, 111
Gross, Gerald, 126
ground station, communications satellite, 1, 30–31; companies interested in market for, 77, 150–51;

National Advisory Committee for Aeronautics
(NACA), 28, 30
National Aeronautics and Space Council (NASC):
Apollo decision and, 71; coordinating government agencies for developing policy for global
satellite communications system, 73, 76, 78,
80–81; drafting policy statement for communications satellites, 82, 86–92; satellite legislation
and, 96, 133; use by Eisenhower, 59–60; Vice
President Johnson as chair, 59, 64; Welsh as
executive secretary, 60, 62, 64; why first focused
on satellite communications, 65. *See also*
Kennedy administration
National Communications System (NCS), 106–7,
122, 128, 134
national security: communications in the Americas and, 122; domestic policy and, 16, 21, 23, 44,
78, 178; IRAC and, 45–47, 120; launchers and,
153; science and technology and, 13–14; Space
Radio Conference and, 17, 105–8, 117; as symbolic and material struggle for hearts and minds,
21–22, 38, 50–51, 56, 143; undersea cables and, 5,
8, 11; US frequency allocation policy and, 116
National Security Action Memorandum No. 342,
180
National Telecommunications Research Center
(CNET), France, 154
Naval Research Laboratory (NRL), 24, 26, 28
neoliberalism, 186
nonaligned countries. *See* Global South
North-South global division, 13–14, 16, 18, 39, 142,
175, 178, 183
NSC-68, 23–24
NSF (National Science Foundation), 5
Nunn, Robert, Jr., 34, 78, 116

O'Brien, Lawrence F., 85
Olympics, Tokyo, 3
Orion, 186–87
Orrick Committee, 106
Orrick, William H., Jr., 106
Our World (television program), 3–4

Pakistan, 136, 151, 185
Palapa A, 185,
PanAmSat, 186–87, 189
Pastore, John O., 45–46, 97, 134, 169
Philco Corporation, 30, 65–66, 73–74

Pierce, John Robinson, 26–27, 30–32
Pierson, Ball & Dowd, 66, 69
pornosat project, 69–70
Potts, James, 111
President's Science Advisory Committee (PSAC),
28, 61
Project Share, 185
propaganda: Cold War and, 13, 56, 62; communications satellites and, 16, 39, 77, 101, 131, 178; direct
broadcast satellites and, 154, 181; Gagarin flight
and, 70; implications for cooperative activities,
170; Sputnik and, 23, 28, 56. *See also* hearts and
minds; soft power
privatization, 186–89. *See also* neoliberalism
PTT (Post, Telegraph and Telephone Service), 65,
70, 155, 169, 174, 186

radio astronomy, 31, 48, 129, 135, 139
radio spectrum: CCIR meeting and, 115–22; early
radio technologies and use of, 9; needs of the
global communications system, 117–18; 1959
ITU meeting and, 47–50; policy in the United
States, 37–47, 75; Space Radio Conference and,
105–43; US military requirements for, 106–9,
115. *See also* International Telecommunication
Union
radiotelegraphy, 8–10
radiotelephony, 9–10, 25, 33
railroad release, 68, 70
RAND Corporation, 23–24, 110
RCA (Radio Corporation of America): beginning
before World War II, 9–10, 12; communications
satellite policy (US) during early 1960s and, 68,
73; competing with Intelsat, 186; former employees at Comsat, 110–12; Intelsat Earth stations
and, 183; Relay project and, 2, 37, 66, 110; *Satcom
1* and, 185; Space Radio Conference and, 119, 131;
telex service and, 67
Reagan, Ronald, 186–87
regionalism, 18–19, 145–46, 155–56, 158, 164–65, 175,
184–86.
Reiger, Siegfried, 110
Relay 1, 2–3, 66, 68, 110
Rogers, William, 34
Rostow, Walt Whitman, 91, 121
Rubel, John, 29, 79, 116
Rusk, Dean, 58, 82, 100
Ryan, William Fitts, 95, 97